P9-DHP-551

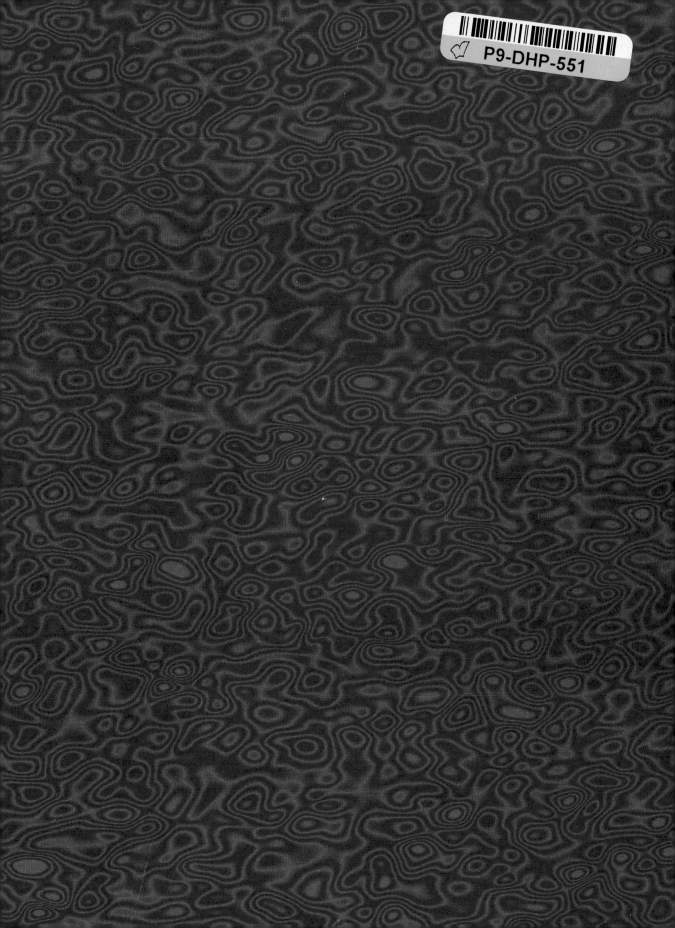

TO THE PUBLIC SEX MEANS NAKED MEN AND WOMEN COUPLING;

TO THE PHYSICIANS SEX IS THE CAUSE OF AIDS;

TO THE MORALIST, CONSTRAINED BY RIGID DICHOTOMIES,

SEX MEANS ONLY TWO ALTERNATIVES: MALE OR FEMALE.

WHY IS SEX SO MISUNDERSTOOD?

IS IT BECAUSE NO ONE KNOWS ITS HISTORY?

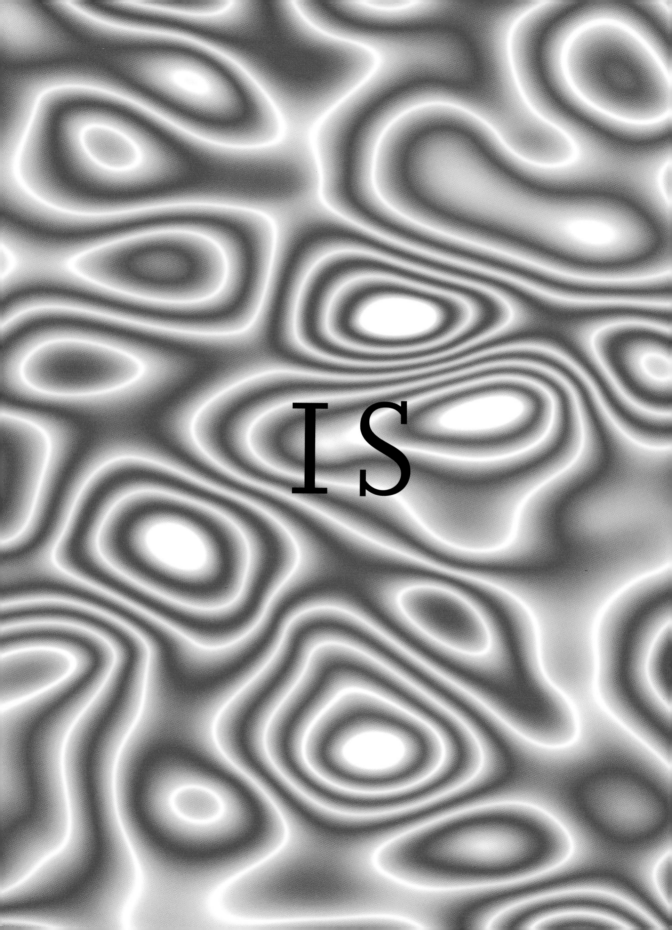

IS

WHY DO WE SCIENTISTS IGNORE THE BASIC ARITHMETIC OF LIFE?

IN BIOLOGY 1+1 DOES NOT EQUAL 2.

RATHER 1+1=1, AS IN ONE SPERM

PLUS ONE EGG EQUALS ONE FERTILIZED EGG.

IS THIS WHY OUR EFFORTS TO PREDICT SEXUAL BEHAVIOR

WITH FANCY COMPUTER MODELS ARE BOUND TO FAIL?

SEX INITIATES ADOLESCENT REBELLION,
JEALOUS RAGE, ROMANTIC FANTASY,
RECKLESS GAMBLES AND NEWBORN BABIES.
WHY IS SEX SUCH A POWERFUL AND
MYSTERIOUS FORCE IN OUR LIVES?

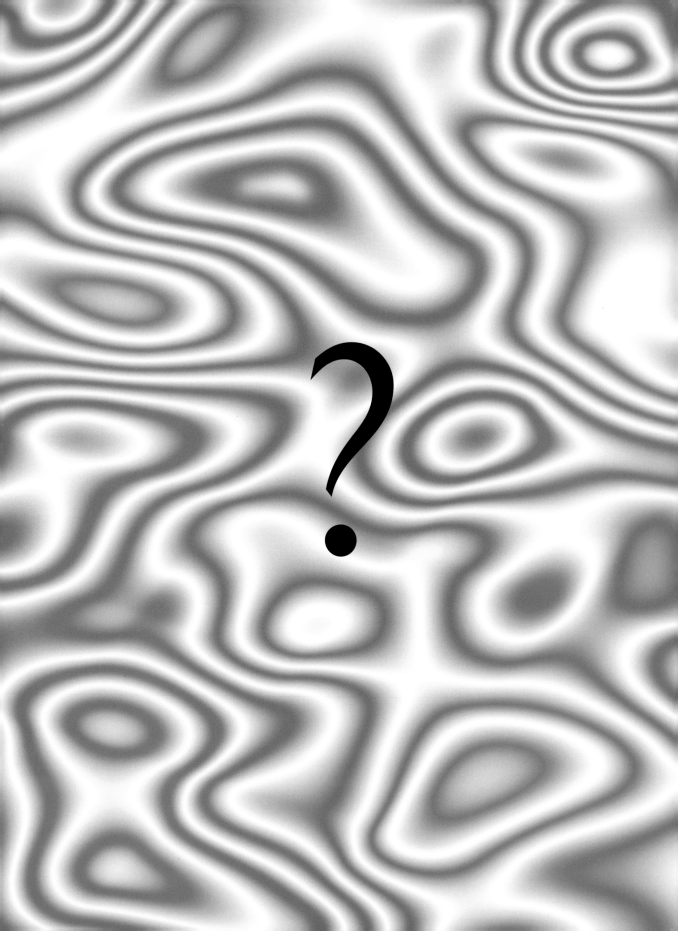

LYNN MARGULIS AND DORION SAGAN

A PETER N. NEVRAUMONT BOOK

SIMON &
SCHUSTER
EDITIONS

Simon & Schuster Editions
Rockefeller Center
1230 Avenue of the Americas
New York, New York 10020

A Peter N. Nevraumont Book

Copyright © 1997 by Lynn Margulis and Dorion Sagan
The credits accompanying illustrations are an extension of
this copyright page.

All rights reserved, including the right of reproduction in
whole or in part in any form.

Simon & Schuster Editions and colophon are trademarks
of Simon & Schuster Inc.

10 9 8 7 6 5 4 3 2 1

Printed in Italy by Editoriale Bortolazzi-Stei, Verona, Italy

Designed by José Conde, Studio Pepín, Tokyo, Japan

Created and Produced by
Nevraumont Publishing Company
New York, New York

Ann J. Perrini, President

Library of Congress Cataloging-in-Publication Data

Margulis, Lynn, date.
        What is sex? / Lynn Margulis and Dorion Sagan:
                p.   cm.
        "A Peter N. Nevraumont Book"
        Includes bibliographical references and index.
        1. Sex—History.   2. Sex (Biology)—History. I. Sagan, Dorion.
        II. Title.
        HQ12.M355   1997
        613.9'5—dc21
                                97-19220
                                CIP

        ISBN 0-684-82691-7

Other books by Lynn Margulis and Dorion Sagan

What Is Life?
Slanted Truths: Essays on Symbiosis, Gaia, and Evolution
Microcosmos: Four Billion Years of Evolution from our Microbial Ancestors
Garden of Microbial Delights
Mystery Dance: On the Evolution of Human Sexuality
Origins of Sex

Other books by Lynn Margulis

Symbiosis in Cell Evolution
Environmental Evolution (with L. Olendzenski)
Symbiosis as a Source of Evolutionary Innovation (with R. Fester)
Five Kingdoms: An Illustrated Guide to the Phyla of Life on Earth (with K.V. Schwartz)
Illustrated Glossary of the Protoctista (with L. Olendzenski and H.I. McKhann)
The Illustrated Five Kingdoms (with K.V. Schwartz and M. Dolan)

Other Books by Dorion Sagan

Biospheres: Metamorphosis of Planet Earth
Into the Cool: The New Thermodynamics of Creative Destruction (with Eric D. Schneider)

contents

one

A UNIVERSE
IN HEAT:
SEXUAL ENERGY

———————

I can resist everything
except temptation.

— *Oscar Wilde*

ETERNAL DELIGHTS | THE UNIVERSE IS A SWELTER of starlight, swamped in excess, consumed by time. Things burn, change, degrade and die. The life which our ancestors detected in every moving thing, whether earthly or celestial, is cosmically rare. But it is cosmic: in the most fundamental sense, all life here or elsewhere, ancient or modern, is conceivable only as a phenomenon of energy flow, of material exchange in a cosmos bathed in the vast energy of stars. Stars—in the case of life on Earth, our Sun—provide the energy for life to work. The basic operation of life is to trap, store and convert starlight into usable energy. In photosynthesis,[*] photons are incorporated, building up bodies and food; they are the principal energy source for those two most basic and natural pleasures, sex and eating.

Sentient life is attracted to sex and food because, by loving and devouring, life maintains and increases itself. Not all species, however, must have sex in order to reproduce. In all who do, sex is a crucial part of the energy transformation process by which, with delight, they maintain and increase their complexities in this energy-steeped cosmos. Moreover, the inevitability of death was foreign to the first living bodies. Living bodies were originally immortal. The death which we fear as end of our individual existence is, as we shall see, intimately related to the evolution, about a billion years ago, of sexually reproducing organisms.

Despite the mortality we inherit as sexually reproducing creatures, sex is great. It produces pleasure and brings babies, the future of humankind, into the world. Without the sex act of our parents, none of us would be here. With sex, each of us is not only a living, breathing, thinking entity, but also a unique mixture of genes from separate sources—in short, a unique individual. The evolution of sex was the greatest boon to individuality the world has ever seen. Although sex does not do so for all organisms, for animals, sex ties us to past places and times because we require it for our reproduction. For its part, the chemical repetition involved in reproduction fulfills a function of energy dissipation and material degradation in

---

* In discussions of sex words become battle cries: A glossary
  of terms, both unfamiliar scientific and familiar ambiguous,
  may be found beginning on page 244.

a cosmos constrained, from our animal perspective, by time's unidirectional flow.

What is sex? It confuses us not only because it literally has to do with the mingling of two distinctly different beings, opening us up to each other in the deepest way, but also because we tend to make mistaken extrapolations about sexuality's importance. Our own biologically parochial existence as sexually reproducing beings does not mean, for instance, that there is only copulatory, genital-based sex or that sex has anything necessarily to do with reproduction. In fact, most members of four out of the five kingdoms of living beings do not require sex for reproduction.[1]*

At the most basic level, sex is genetic recombination. It is the mixing or union of genes, that is DNA molecules from more than one source. [PLATE 1] When a DNA molecule makes another DNA molecule just like itself, biologists speak of replication. When, however, living matter, as a cell or composite body of cells, produces another similar living being, scientists speak of reproduction. [PLATE 2] The broad, biological meaning of sex simply refers to the recombination of genes from separate sources to produce a new individual. Sex is not equivalent to reproduction. On the one hand, any organism can receive new genes—can indulge in sex—without reproducing itself. On the other hand, plants bud, bacteria divide and cells with nuclei reproduce all without any requirement for sex. Such sexless reproducers include amebas and even the constituent cells of your body. We associate sex with reproduction not because they are necessarily or logically linked but because they became linked contingently during the specific evolution of our animal ancestors. As we shall see by the last chapter, the unlinking of sex from reproduction, via cybersex and contraceptive technology, is part of the shape of things to come.

Sex involves the acquisition of new genes, a shuffling of genetic information that sometimes leads, as in a game of cards, to a more effective combination—the biological equivalent of a better hand. Sex, as we know

---

Endnotes can be found starting on page 229.

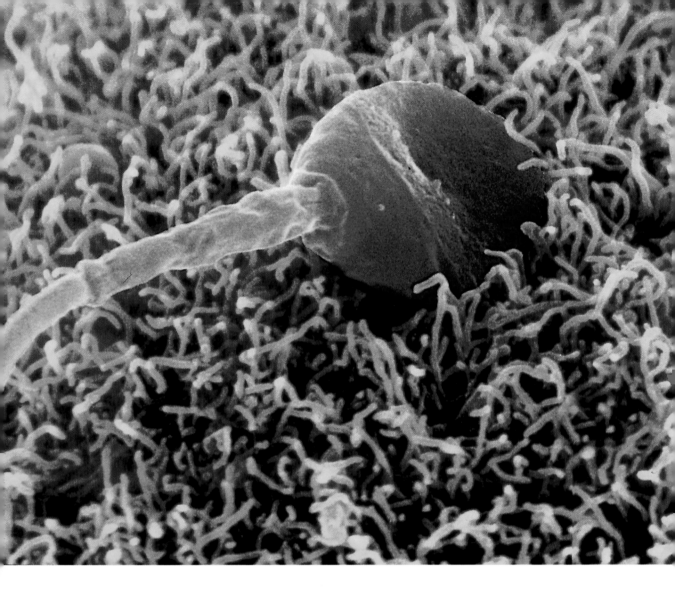

PLATE 1
**Sperm from a man's ejaculate attempting
to fertilize a woman's egg. Color enhanced
scanning electron micrograph.**
[David Phillips/Photo Researchers]

by comparing ourselves to our parents, produces variation. But plenty of other processes—such as cosmic irradiation, acquisition of viruses or symbionts, or exposure to ambient chemicals—also alter the structure of DNA or add DNA to produce variation. When we look at life in its cosmic setting, what is striking is not so much life's production of variation but its ability to produce near-perfect copies of itself. In sexually reproducing species, which may number up to tens of millions, sex itself is the means by which organisms produce such near-perfect copies. Organisms are able to recognize members of their own species and choose members of the opposite gender[2] by using very subtle cues. Whether male or female, whether you have brown, blonde or red hair, you are—from the broad, cosmic perspective—almost exactly like your parents. Sexually reproducing or not, organisms pass their identity on to their offspring with remarkably few changes.

The strong human interest in sex is directly related to its status as a key part of our life history. Our ancestors have been sexual reproducers for perhaps 600 million years, since the origin of the animals. To understand our passionate interest in sex, you have to understand the role of sex both in reproduction and apart from it. Aside from reproduction sex is, as we shall see, part of a natural tendency to mix things up, to randomize, to lose discrete identity due to the tendency of material systems to reach more probable states. But, in us, sex is intricately tied to reproduction, and as such, it takes on a different aspect, one that has as much to do with preserving identity as with destroying it. Ultimately, sexual reproduction is the fundamental biological process of maintaining and reproducing identity.

Although the reproductive process of making living copies might seem to be life's most fundamental trait, it is in fact a secondary one. Reproduction is based on autopoiesis. From Greek words meaning "self-making," autopoiesis is the fundamental property of living things. Reproduction, making others like yourself, is conceptually an offshoot of keeping yourself as you are. Unlike inert objects, living things are exposed to continuous

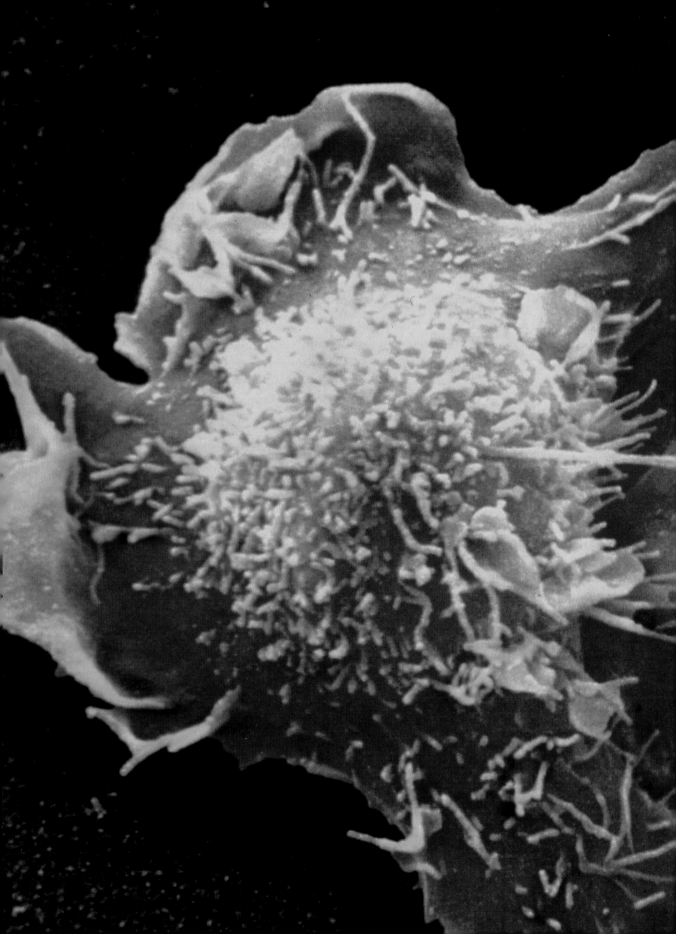

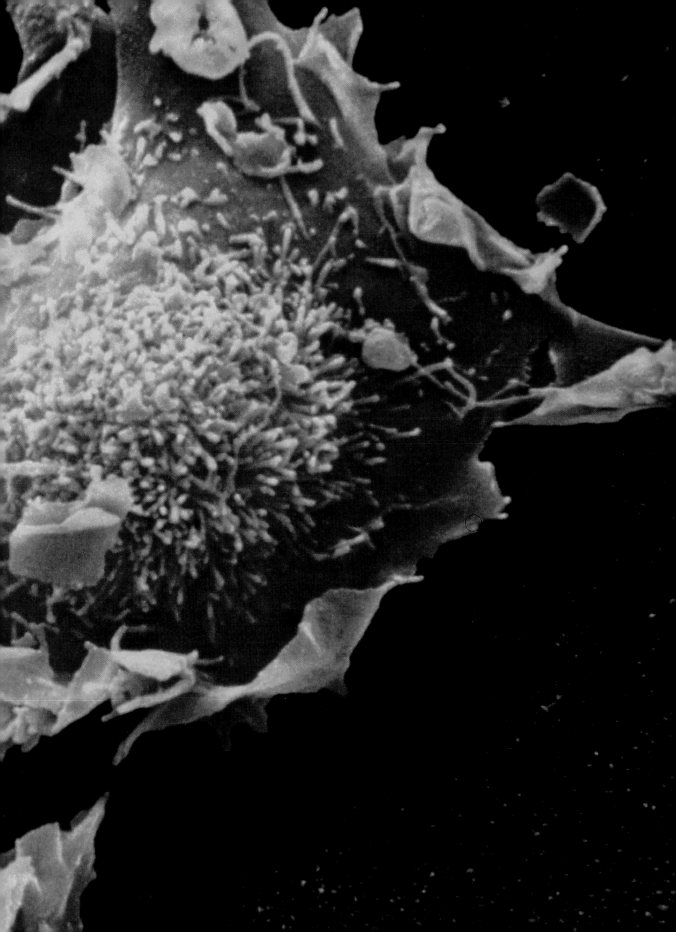

material and energetic flows. Employing energy to run our numerous bio-chemical processes, autopoietic networks—live beings—continually recycle their components to maintain themselves. They metabolize. The ability to change to stay the same, to employ energy flux to power the cyclical turnover of matter necessary to maintain a self, is the basic biochemical trick of autopoiesis.

According to biologist-philosopher Gail Fleischaker, all autopoietic systems share three traits: they are self-bounded, self-generating and self-perpetuating. They are self-bounded in that autopoietic systems are sur-rounded by a cell membrane, skin or shell that simultaneously encloses the system while allowing it continuity with energy and materials from the outside world. They are self-generating, such that the entire system, includ-ing the boundary, is produced by the system itself. Finally, they are self-per-petuating, meaning that autopoietic systems, even if they are not growing or reproducing, use energy continuously to maintain their relatively com-plex forms. Humberto Maturana, the Chilean biologist who invented the notion of autopoiesis, points out that the behavior of a living organism is, like other natural objects, determined. The difference is that the living being is determined largely by its own internal processes rather than by any outside force. A billiard ball, that paradigmatic object of classic Newtonian physics, can only react. We living beings, referring not only to the outside world but also to our own selves, can also act. We have a freedom, a self-referential complexity, not found in inert objects. This freedom and com-plexity is the result of our autopoietic closure, a separateness from the out-side world more fundamentally a trait of life even than the commonly not-ed property of reproduction. Autopoiesis is like a successful restaurant, and growth and reproduction are like the expansion and franchising of that restaurant.

But the aloofness and identity of living systems is partial. We are con-nected to the cosmos. The autopoietic prerequisite to reproduction, includ-

PLATE 2 *preceding page*
**Offspring cells of a recent reproduction,
still joined by a slender strand.**
[*Keith Porter/Photo Researchers*]

ing sexual reproduction, depends crucially on fluxes of energy. The islands we call life are conceivable only in the context of a cosmic ocean of energy transformation. In the early 20th century, the Russian scientist Vladimir Vernadsky (1863-1945), who popularized the term biosphere, depicted living matter as "green fire" that could be understood only in terms of a transformation of the energy of the Sun. The Sun's energy, "radiant and chemical, working in conjunction with the energy of chemical elements, is the primary source of the creation of living matter."[3]

Influenced by his teacher, V.V. Dokuchaev, who proposed that soil was not only a geological but a biological phenomenon, Vernadsky expanded a biological view of the physical Earth to encompass the entire surface layer of life. Eschewing the term life, with its theological, philosophical and historical connotations, Vernadsky, a disciplined materialist, always spoke of living matter. World War I, with its transport of munitions, planes and troops, gave Vernadsky the idea that human activity could also be looked at as a geological phenomenon. The energy of growth and reproduction and the energy of living beings to move and act was biogeochemical energy, ultimately a transmutation of sunlight. Fossil fuels, for example, were laid down in the Carboniferous period by photosynthesizers such as giant seed ferns (Cycadofilicales) that had autopoietically trapped sunlight and turned it into carbon-based living matter. The energies of life, including the energy for its evolution and the colonization of the biosphere, were solar. Although common in the universe, the basic atoms of living matter—carbon, hydrogen, oxygen, sulfur, nitrogen, and phosphorus—take on a peculiar energy-rich organization on the surface of our planet.

Animals, with their complex tissues and organs, and inevitably aging bodies, evolved from sexually reproducing colonies of microbes called protoctists. Protoctists, in turn, evolved from deep intimacies among very different kinds of bacteria. The bacterial intimacies that led, through protoctists, to the first animals were more than sexual. Very different organisms fused. They began to share the same body, not for a few moments, but forever. Today, inside nearly all of the cells that make your tissues are cell parts, called mitochondria, inherited only from your mother. About two billion years ago their promiscuous ancestors, in a process reminiscent of both sex and infection, penetrated and began living in larger cells. Luckily for the progeny of such beings, they never withdrew but became permanent partners of all plant, animal and protoctist cells.[4]

According to most cosmologists, nuclear physicists, astronomers and space scientists, the universe began in a singularity—the explosion of everything from an immensely hot, infinitely dense point appearing some 13.5 billion years ago. One second after it originated, matter from the "Big Bang" had spread outward to the immense distance of three light-years. It was still too hot for atoms to exist. After three minutes of explosion subatomic particles had settled down to a "cool" billion degrees centigrade and traversed some forty light-years. When scientists observe the so-called red shift—the change in light-wave patterns of stars—they find that today's galaxies are still flying away from each other at immense speeds. The Big Bang continues. In all directions in the sky one finds low-level microwave radiation. This so-called background radiation is the distant "echo" of the giant explosion that started it all. Some of the heavier elements that would eventually wend their way into living matter, such as carbon, were formed only later, baked from lighter elements by a real alchemy in the natural nuclear ovens of stars which later exploded. We are the offspring of particle bombardments and interactions, of pre-sex mergings and more-than-human violence.

The Milky Way, the spiral galaxy upon whose remote edge our solar system dwells, provided the cosmic detritus, the nebulous material from which the particles that were to be the Sun gravitationally cohered. Hydrogen, the most abundant element in the universe, provided the raw material for the powerful nuclear reaction called fusion whose blinding radiation—sunlight—has been worshipped by many human cultures. If a little more massive, hydrogen-rich Jupiter, the most massive planet in our solar system, would also have become a second Sun, making ours a two-star solar system. Earth, still in a molten state, formed in orbit around the Sun about 5 billion years ago. The most ancient solid rocks known in the solar system, from comets, meteorites and the Moon, are some 4.6 billion years old. Current astronomic thought suggests that the Moon itself was produced when a powerful impact withdrew a piece of Earth into orbit around itself. Genetic, fossil and comparative morphological evidence strongly suggest that all life evolved on Earth soon after it had cooled to form a solid crust. Because of its light weight hydrogen tends to escape into space. The flight of hydrogen gas has been thwarted by life. By autopoietic and reproductive recycling Earth has preserved a hydrogen-rich lively environment—a place full of bodies. Our hydrogen-retaining watery Earth loaded with surface carbon, phosphate and sulfur is unique in the solar system. Our living Earth really began to differ from our neighbor planets some four billion years ago when life originated.

Today, the cosmic context of life's origins continues to exert a powerful influence over our lives, including our sex lives. Melatonin, a hormone involved in the timing of breeding, triggers seasonal lust in many animals, including ourselves. Its release from the pineal gland in the brain is stimulated by sunlight. Pacific grunion mate on shore in the full moonlight of spring tide. Although astrology is nonsense, planetary and lunar motions, the cycles and lights of nights and seasons, and especially solar radiation continue to cue and influence the lives of organisms on Earth's surface. Biologists who study biological rhythmicity distinguish exogenous (externally triggered) from endogenous (internally triggered) rhythms. Over time the natural rhythms triggered by events in the outside world have evolved into internal biological clocks as living beings have become increasingly independent of the environment in which they evolved.

All of us dwell in a pervasive, if neglected, cosmic context. *Gonyaulax*, a luminescent protoctist studied by daily (circadian) rhythm expert J. Woodie Hastings of Harvard University, glows in the dark each night like clockwork. Even if sequestered in a locked laboratory away from obvious external cues such as nightfall and sunrise, *Gonyaulax* remembers to glow. With all the playful intrigue of a musical prodigy, living matter has tended, over the four-odd billion years of its evolution, to refashion the clock-like rhythms of the cosmic outside into its own increasingly independent timers. A deeply rhythmic genetic alarm clock generates the changes of puberty, ushers in the urges and defenses of motherhood, and triggers the changes of menopause. The softness of sundown and the cover of night predisposes cyclical sexual beings to bring forth music, dance and revelry— preludes to mating in so many cultures. The theme of living matter internalizing, with increasing variation, the cyclicity of its cosmic surroundings applies also to the rhythm of sexual love as a permutation of the primordial music of the universe.

Musicality returns us to the importance of energy in understanding life and sex. Although not nearly so heavily popularized as chaos theory, fractals, Boolean networks and other mathematical, computer-based simulations of complex patterns, the science of energy flows—thermodynamics— provides us with a developing theoretic perspective and fascinating examples of the emergence of complexity. Moreover, the complex structures studied by thermodynamics are not programmed patterns on a computer screen but real, three-dimensional structures cropping up naturally in the physical world of everyday reality.

Because the sciences of complexity are, above all, interested in modeling the emergence and evolution of life and intelligence, it would seem that the flow patterns studied by thermodynamics deserve greater attention. Indeed, whereas the near-paranoiac vision of the origin of life from a computer program might be accepted for the sake of passing entertainment

---

PLATE 3
Spontaneously appearing Benard convection cells arise due to heat energy influx.
[*Scott Camazine/Photo Researchers*]

between the pages of a science fiction novella, the origin of life from thermodynamic flow structures is virtually assured. A number of observations and experiments support the notion that energy can flow through and organize structures to be more complex than their surroundings, as we shall see. In heated, circular pans of water or silicone oil, fluid hexagonal shapes maintained by convection currents, called Bénard cells, for example, appear spontaneously. [PLATE 3] These shapes emerge on their own from the surrounding disorder. They are organized. Such patterns of beauty are the last thing you would expect to result from random collisions of atoms. The mere existence of autopoietic-like networks which maintain their identity despite or, rather, because of the energy flowing through and feeding their complexity deserves greater scientific scrutiny. Thermodynamic structures, three-dimensional systems that cycle chemicals and energy through themselves maintaining their complexity over time, include more than life. Life is but one example of a wider class of energy-material flow structures that generate complex cyclicity. But being alive ourselves life is the most interesting of these complexifying structures that self-organize due in part to the energy flowing through them.

SENSUOUS FLOWS | The scientific cognizance of the permutations of energy—discounting a few apt generalizations such as Greek philosopher Heraclitus's "everything flows"—began in earnest with the development of thermodynamics in the 19th century. Of note are a few historical threads leading to the modern study of heat flow. In the 16th century, Galileo invented the first known crude thermometer, and in the 17th century, English chemist Robert Boyle (1627-1691) was the first to collect and experiment with gases. He showed that air was compressible, indicating that air is, in fact, composed of particles with space between them. Transforming alchemy to chemistry in his 1661 *The Skeptical Chemist*, Boyle showed that gas volume is inversely proportional to both pressure and temperature (Boyle's

Law). Boyle knew that particles of gas cannot each be individually studied. Yet their behavior can be predicted using methods of statistical sampling. By considering gases as aggregates of particles rather than as individual particles, Boyle made a conceptual break with the strict determinism of Newtonian dynamics.

French physicist Nicolas Leonard Sadi Carnot (1796-1832), trying to improve the steam engine, found that maximum efficiency depended upon the temperature difference within the mechanism. The first to quantify the relationship between heat and work, he is considered the founder of modern thermodynamics. His discoveries that energy is conserved, yet that it is impossible to turn all heat into work, are still among the clearest statements of what we now know as the First and Second Laws of Thermodynamics.

The First Law of Thermodynamics concerns quantity: in a closed system, the total quantity of energy, whatever its transformations, will remain unchanged. The Second Law of Thermodynamics concerns quality: in a closed system, high-quality energy is inevitably lost to friction in the form of heat. Carnot's realization of an inevitable erosion of the quality of energy was the first to put humanity on notice, from within science, that the universe is not symmetrical with regard to time. Complex processes, including those of life, have tendencies and directions. As we shall see, the evolutionary trend toward complexity, including human sexual love, with its multi-million year history, is probably undergirded by the thermodynamics-based asymmetry of time.

Although since proven wrong, Carnot believed that heat was an invisible fluid. He believed that the hot-to-cold "downhill" motion of heat—like a waterfall flipping a waterwheel—was the source of energy. Like many chemists of that period, Carnot borrowed his notion of heat from the French chemist Antoine Laurent Lavoisier (1743-1794) who showed that air was composed primarily of two distinct gases—combustible oxygen and noncombustible nitrogen—and whose careful measurements set the tone for modern chemistry. The 18th century Scottish scientist James Black also treated heat as an invisible fluid—"calor." His term calorie, the quantity needed to raise one pound of water one degree Fahrenheit, is still in use, although his conception of heat as a fluid has been superseded by the modern view that heat is a result of the movement of atoms.

A famous thought experiment represents the shift from the ancient, deterministic view of heat as flowing substance to the modern view—called statistical mechanics—of heat as the result of probabilistic atomic interactions. In 1871, Scottish physicist James Clerk Maxwell (1831-1879) suggested that a tiny demon, guarding a door between two compartments of equal temperature, might let only fast moving particles from one side through to raise the temperature of the other side. Thus it was not true, as in the earlier perspective, that heat was a substance that could only flow from a warm body to a cold one. Under certain, albeit highly improbable, circumstances, a cool body next to a hot one might get cooler. Even before quantum mechanics, thermodynamics chopped away at the edifice of Newtonian determinism, replacing inevitability with probability.

In the same century, Joseph Louis Gay-Lussac (1778-1850) showed that gas pressure increases (or decreases) by 1/273 of its initial value for each degree Centigrade increase (or decrease). Thus in theory at -273 degrees Centigrade—that is, at 0 degrees Kelvin or "absolute zero"—all molecular action is predicted to cease as the gas squeezes into zero volume. Nonetheless, this extrapolation, later considered part of the Third Law of Thermodynamics, has still not been experimentally verified due to the inherent technical difficulties of reaching such a low temperature.

Classical or equilibrium thermodynamics—from the early heat flow observations to the probability-based statistical mechanics of atoms—studied systems closed to energy flow. German physicist Rudolf Clausius (1822-1888) introduced the term "entropy" as a measure of the one-way conversion of energy into heat and friction in such a closed system. Basing his mathematics partially on the statistical mechanics of Scottish physicist James Clerk Maxwell, Austrian physicist Ludwig Boltzmann (1844-1906) later explained the one-way nature of this conversion by showing that, in a mapping of gas particles distributed into two chambers, there were far more disordered states (i.e., a mixture of particles in various states) than ordered states (i.e., a mixture of particles in a limited number of states). There were, in other words, many more ways for particles to be distributed evenly than lopsidedly. Probability was on the side of disorder, mixing and dissipation. The famous Second Law of Thermodynamics, the Grim Reaper of nature, states that disorder (entropy) in any closed system must increase. The probable state of particles is one in which their energy, unconcentrated, is of little use. Heat, for example, is useless relative to the sunlight that generates it.

There are critics even of this classical thermodynamic view. Huw Price, an Australian philosopher knowledgeable in physics, argues that the asymmetrical view of time, inherent in thermodynamics, is inconsistent with classical Newtonian mechanics and should be abandoned. Price points out that the probability of future states of disorder is—from a classical, time-symmetric view of physics—matched by the probability of disorder in the past. According to one popular version of modern cosmology ("the inflationary universe model"), matter was distributed extremely smoothly (i.e., ordered) shortly after the Big Bang. Yet, according to theories of how the force of gravity acts upon matter (for example, in the stellar collapses known as black holes), the most probable state for the early universe would have been much more "clumpy" (i.e., disordered). If the distribution of matter in early universe was smooth, as the microwave background radiation suggests, what permitted it to display such great improbability? If Price is correct and the laws of physics are symmetrical with respect to time, the cosmos should become more probable and disorderly no matter in which direction we move.

Nonetheless, we are born, sexually reproduce, and die rather than *vice versa*. It is difficult, if not impossible, to conceive of evolution, let alone daily life, without a temporal direction. The debate is very old, as attested to by the difference between the view of the pre-Socratic Greek philosopher Heraclitus (about 540-475 BC), who held that everything changes, and Parmenides (about 515-450 BC), who believed being and reason to be real while the act of becoming, which is perceived by our senses, to be an illusion. Zeno, whose paradoxes prove by logic that motion does not exist, was a student of Parmenides. Plato, who devoted a dialogue to Parmenides, also held that ultimate reality is eternal, whereas change, as we perceive it on Earth, is an imperfect shadow of the timeless realm of ideas. But these criticisms of temporal asymmetry do not seem to apply to the real world of life where nonequilibrium thermodynamics presents us with a most comprehensive theoretical backdrop.

There are two major differences between classical thermodynamics and nonequilibrium thermodynamics. The first is that classical thermodynamics studies structures of decreasing complexity—machines that lose the capacity for work—whereas nonequilibrium thermodynamics studies entities, including living beings, which increase their complexity and gain a capacity for work. The second difference, fundamentally related to the first, is that classical thermodynamics studies closed and isolated systems while nonequilibrium thermodynamics focuses on open systems. Closed systems are sealed to incoming matter. Matter, by contrast, flows through open systems. In a living body, to take a prime example, matter enters the system as food, drink, and air and then, after transformation, residua are excreted. As University of Georgia ecologist Eugene Odum said, of open systems (he was talking about life), "matter circulates, energy dissipates."[5]

Except for incoming meteorites, the complex system of life on Earth—the biosphere—is a closed system: cosmic rays and solar radiation enter the system, but matter in general does not. Individual organisms by contrast are open to both energy and materials flowing through them. Indeed, the most basic parts of living—eating, breathing, excreting, sex—attest to our status as open thermodynamic systems. It is probably no coincidence that the most natural pleasures—such as thrusting, coming, sneezing, drinking, eating, defecating, urinating, sunbathing, sweating, and even music and vision as the aesthetic delights of sound entering the ear or light waves dancing through the black holes of our pupils to create visual impressions at the back of our retinas—tend to involve orifices and flows.

The self is a self because of its informational closure—we think of and refer to ourselves as discrete entities. We name ourselves and then decorate our names with identifying numbers and titles (doctor, minister, attorney-at-law, professor, etc.) to attest to our relative independence and separation from one another. Such closure is exacerbated by the American ethos of individualism. But this ethos tends to prejudice our perception against the basic biological reality that we are open systems whose very existence depends upon the flow of energy and matter through us. Moreover, in our sexual activity, we are not only thermodynamically, but informationally, open: our existence (although not that of all organisms) depends upon the combining of DNA from one source-parent to DNA from another source-parent. Thus, we are not only energetically and materially open, but we are also informationally open. In order to continue evolutionarily we open ourselves to fresh genes.

Beyond open and closed, thermodynamics also defines isolated systems. Isolated systems are closed, not only to material influx and outflow, but also to the leakage or entry of energy from or into the system. Without an energy source with which to work, these so-called adiabatic systems, upon which much classical thermodynamics is based, tend to come to a standstill. Extrapolation of the tendency for isolated systems to "die out" led to the notion of the heat death of the universe: the idea that the entire universe will react until all the stars are burnt out and it becomes a lifeless, boring, uniform-temperature cosmic wasteland. As, however, astrophysicist Freeman Dyson has pointed out, this conclusion is cosmologically premature. Stars themselves are far from equilibrium, despite 15 billion years of cosmic evolution. What we call the universe may itself be part of a larger open system. The evidence needed to extrapolate our cosmic destiny remains incomplete.

Nonequilibrium thermodynamics was the first science of complexity. A key figure in nonequilibrium thermodynamics is the Russian-Belgian chemist Ilya Prigogine who was awarded the Nobel Prize for mathematically describing the behavior of heat convection cells, long-lived chemical reactions called chemical clocks and other naturally complexifying non-living systems he classified as "dissipative structures." As Prigogine points out, these systems exhibit a kind of memory. Their mature form strongly depends on the initial conditions. In this they resemble the energy-dissipating systems of life. But the "life span" of Prigogine's dissipative structures, measured in hours, is blinkingly brief compared to bacterial cells, sexually cavorting protoctists and animal bodies—all of which are part of the single, continuously cycling dissipative system of life on Earth that has continued for over 3.5 billion years—the biosphere.

Entropy in open systems, such as those of life, is notoriously difficult to measure. Nonetheless, it appears that the tendency for time to display a direction is behind Prigogine's dissipative structures. Dissipation, depletion and languishing toward stasis—the asymmetric direction formalized in the Second Law—may be the universe's destination but, paradoxically, in the process of getting there it is capable of producing structures of increasing, rather than decreasing complexity. Indeed, where and when energy flows structures of surprising orderliness sometime spontaneously emerge. These structures, nonetheless, ultimately function to produce disorder and thus obey the Second Law. The overwhelming ease of burning a book, trashing a room or breaking apart a jigsaw puzzle relative to authoring, constructing or organizing reminds us of how even orderliness generates disorder. This becomes obvious when we consider how much mess, garbage and pollution, how much food and excrement the average animal—let alone industrial civilization—leaves in its wake. Local order, as Prigogine stresses, necessarily produces disorder further out. No process, not even life, is exempt from the Second Law.

Thermodynamicist Harold Morowitz, director of the new History of Consciousness program at George Mason University, explains the Second Law like this:

> The tendency of energy to become randomly distributed in the kinetic form among molecules is the basis of the famous Second Law of Thermodynamics. There are almost as many statements of this law as there are thermodynamicists, but all of them convey the idea that in an isolated system entropy will increase. Thus heat flows from a hotter to a colder body, and molecules diffuse from a higher to a lower concentration. A system moving toward equilibrium assumes the most disordered molecular state consistent with the conditions under which it is maintained. At equilibrium everything is completely homogeneous and nothing interesting can happen.[6]

The Second Law which, because of its emphasis on increasing disorder, seems to contradict the evolution of complexity in living matter, in fact provides a deeper explanation for life's behavior. Life's complexity is matched by an equal if not greater tendency to produce disorder—heat and entropy—around it. Sex, an increasingly intimate part of the processes by which living matter perpetuates itself, has a dual relationship to the increase in disorder mandated by the Second Law. On the one hand, sex has become a mandatory facet of reproduction for many organisms, including humans and other familiar animals. Producing the order of living copies, by making self-similar living matter, sexual reproduction accelerates the production of disorder because the complex natural organization of a living being inevitably exports more disorder to pay for its organization. On the other hand, producing self-similar living matter in a universe heading toward disorder will never be perfect. Sex, embedded in animal reproduction, exemplifies the natural tendency for all copying processes to be imperfect. Therefore, sex both protects and disturbs the integrity of living matter, and yet does both in accord with the Second Law of Thermodynamics—a principle that not only governs nonliving and living matter, but helps us explain why living matter exists and behaves as it does.

THE SCHRÖDINGER PARADOX | If the Second Law of Thermodynamics teaches a universal increase toward disorder, why does the evolution of life—with its equally impressive general increase toward order—apparently violate it? This paradox, posed by Austrian scientist Erwin Schrödinger in his 1943 Cambridge University lectures, drew attention to biology as a problem of physics. Much has been made of Schrödinger's inference that some informational molecule must be responsible for heredity. He deduced that life acted as an aperiodic crystal.[8] His writings inspired scientists to seek the material basis of heredity. Ten years after Schrödinger's book, James D. Watson and Francis Crick announced their discovery of DNA's double-stranded, helical structure. Identifying the molecule capable of self-replication allowed others to show how DNA, in combination with RNA and protein, directed the maintenance, growth and reproduction of cells. Yet an equally serious problem, the potential rapprochement between biology and physics, remained unanswered: How do organisms manage to resist thermodynamic decay? Somehow, Schrödinger conjectured, living beings must concentrate a stream of order upon themselves, a stream to which Schrödinger gave the fanciful name of "negentropy."

Classical Newtonian mechanics has no privileged direction in time. The equations describing the motions of billiard balls bouncing off of each other are equally adept at describing the billiard balls in a reversed film where they travel backwards.[9] Nonetheless, in our real-life experience, time moves in only one direction—from the past to the future. Cars depreciate, people age and flowers bloom and die each sweet-smelling spring: in the real world time has direction. Cream plopped in coffee forms clouds that merge to form a beige mixture. Coffee cools; it never spontaneously warms itself up or segregates back into its black coffee and white cream components. A lit match flares, smoke curling, sulfur dissipating throughout the room. Only in a backwards-projected film do we see smoke converging mysteriously onto a matchhead, spilled milk returning to the pitcher, or a broken champagne glass perfectly reassembling itself in midair as it travels against gravity to the hostess's outstretched hand. But such miraculous improbabilities show precisely the stream of order Schrödinger detected working forward in time in life and which he called upon his fellow scientists to explain.

Schrödinger first suggested an answer by the way in which he posed the question. Life constructs its cells, bodies and brains by using DNA as a template or blueprint to assemble the carbon-, hydrogen-, phosphorus- and sulfur-rich matter of cells. Watery matter moves from its not-so-ordered environment into the DNA-dominated domain of its highly ordered self. But since life, as Schrödinger stressed, is not exempt from the laws of physics and chemistry, there must be a price to pay for this order. Life cannot violate the Second Law. The order of life comes from the Sun.

Imagine that your favorite potted philodendron disappeared into a stream of light aimed at this Sun. This image—your house plant as perceived in the opposite direction of time—shows that organisms are not isolated; they are open systems embedded in an environment. The sum total of life, which today derives little matter from outer space, is non-negotiably open to a continuous energy flux from the Sun. Life's basic process is to take the Sun's low-entropy, long-wavelength photons of visible and ultraviolet light and reradiate them as shorter-wavelength infrared radiation—in other words, life converts light to living matter and heat. Trapping, using and, to a certain extent, recycling the high-quality energy of photons coming from the Sun, life lives and grows, producing entropy and heat as its cosmic waste. Life appropriates these photons, takes them out of circulation for a while, but then returns them to outer space as heat. If life were an isolated system, it would be miraculous. But it is not, it is an open system. Evolution's rise in complexity is funded by low-entropy solar radiation. Even the dream of the sleeping tiger, digesting a gazelle whose own body comes from grazing plants, is funded by the highly ordered energy of the Sun. Far from being a kind of independent perpetual motion machine in defiance of the laws of physics, life's order is a borrowing or spending of enery from the Sun.

In the cosmic scheme of things, the increase over evolutionary time of the complexity of living matter on Earth is perfectly natural. Life is not unique but only the most impressive example in a larger class of entropy-producing structures. These dissipative structures increase their order locally by dissipating the energy flowing through them, energy which adds to an overall increase in disorder. Like borrowing money that must be paid back with interest, life's order, as that of all dissipative systems, is not had for free but at the cost of greater disorder.

If we take a step back and look at life from a larger viewpoint that includes not only organisms but their environment, we find many ways in which they create disorder. Urine, feces, sweat, pollution, garbage, and carbon dioxide exhalation are all examples of the inevitable messes made by human life. As open systems we must dump our material and energetic wastes. At the level of the organism, we do so by exhaling and excreting energy-poor gases, fluids and feces. At the ecosystem level, we excrete by dumping sewage and creating landfills around our cities. On a biospheric level, we inevitably pollute the global commons of the world's oceans and hurl rocket leftovers into the near-Earth orbits of space. By burning fossil fuels, for example, human beings pollute Earth's atmosphere with the thermodynamically more probable waste product carbon dioxide. Although life, by nature, is an adaptable, evolving system—one so subtle and able it has found ways of recycling wastes back into living matter—it cannot escape the thermodynamic mandate that local order creates global waste. Trees, for example, can recycle limited amounts of carbon dioxide. But the fact that industrial civilization has produced so much carbon dioxide—by far the most abundant gas in the atmospheres of our lifeless planetary neighbors Mars and Venus—is sobering testimony to the dictates of the Second Law.

A little knowledge can be a dangerous thing. In the 1960s and 1970s, some writers made apocalyptic pronouncements on the inevitable decline of civilization based on a superficial understanding of the tendency toward disorder implicit in the Second Law. But the Second Law only states that entropy will rise in isolated systems, and Earth is not isolated. Life must be open to the Sun and to its environment of astronomical cycles as all beachlovers and anyone afflicted by seasonal depression or spring fever knows. The complexity of living matter—including that part of its complexity which is dependent upon sexuality—can increase as long as the Sun is there to power it.

Perhaps the clearest scientific evidence that life obeys the Second Law concerns temperature. Satellite observations show that habitats on Earth's surface which are rich in life are the most adept at cooling—and thus producing heat as entropy above them in accordance with the Second Law. Indeed, mature, biodiverse ecosystems such as the Amazon rain forest, with its millions of highly integrated sexual species, are the best coolers of all. Pumping water through tree roots, which evaporates at the surface of leaves in a process known as evapotranspiration, rain forests are effective natural air conditioning systems whose local complexity is more than compensated for by the heat they generate into space. Interestingly, volatile compounds called isoprenoids, released through openings in the leaves of trees

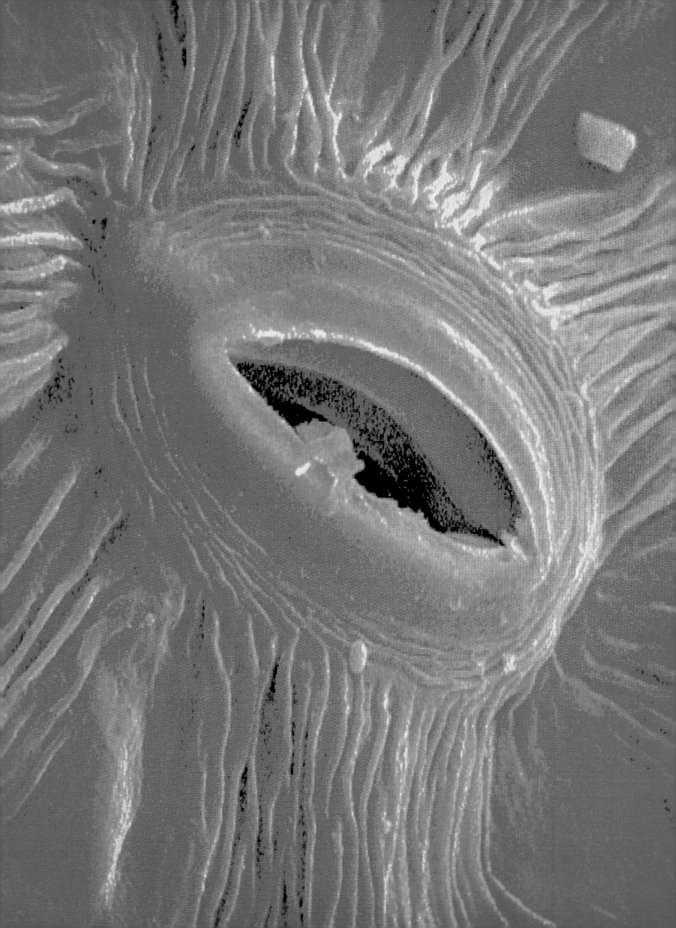

called stoma, stimulate the precipitation of rain. [**PLATE 4**] The isoprenoids serve as nuclei for the condensation of rain droplets. Thus the rain forest and rain are a single locally cooling, global entropy-producing system. The system really extends to the "inanimate" weather above the canopy as well. From a physical standpoint, then, living matter is part of an open thermodynamic system whose effects cannot be understood without looking beyond the surgical closure of skin, bark or shell. Life and sex make sense only in the context of an energetic universe.

Moreover, fossil evidence suggests that Earth's surface has been cooler than had no life evolved. Life, inhabiting a temperature range between the freezing and boiling points of water, has a record of continuous presence on this planet for over three billion years, indicating relatively moderate temperatures. According to models of stellar evolution, however, the Sun has steadily increased in temperature. The best indications suggest that Earth, with life, has cooled itself off, whereas if Earth had been devoid of life it would have had a hotter surface or local temperature yet have had little capacity to produce heat and entropy globally. Just like a refrigerator, which adds heat to your kitchen despite its inner coolness, so life, despite or because of its cooling abilities, adds heat to its surroundings. Earth cools itself in a number of ways which are under active investigation by the international scientific community. The duration of Earth cooling by forest evapotranspiration and burial of heat-trapping "greenhouse gases" may be naturally limited as life runs out of ways to battle the Sun's increasing luminosity. The ultimate expansion of the Sun into a red giant is predicted to turn Earth's ocean water into steam. Yet by the time the Sun is gone, some five or so billion years from now, who is to say that life will not have expanded to fill niches on planets of other stars, growing and breeding around new suns?

There are almost as many statements of the Second Law as there are

PLATE 4
Stomate (breathing pore) from the underside of a leaf of an elder tree (*Sambucus nigra*). Stomates are organs of gas exchange between the leaf interior and the atmosphere. These pressure-regulated pores are generally open (as shown here) by day and closed at night.
[*Andrew Syred / Science Photo Library*]

thermodynamicists, as Morowitz said. Perhaps the most useful one for open systems is the version coined by Eric D. Schneider, a thermodynamicist at the Hawkwood Institute in Livingston, Montana. Schneider's assertion, that nature tends to reduce gradients—differences of all kinds, not only those of temperature required to do work, but of pressure, chemical concentration, and so on—is an elegant extension of the Second Law. Nature abhorring the reduced pressure in a vacuum flask or the cool of ice on a hot day spontaneously evens out these differences.

The convection cells shown in **PLATE 3** are the result of reducing the gradient between the container's cool top and hot bottom. A highly complex weather system, such as a tornado, has a characteristic shape which would never be predicted on the basis of equilibrium thermodynamics or probability alone. [**PLATE 5**] The sides of a tornado or a whirlpool are almost vertical. Certainly the system—not at rest at all—is out of gravitational equilibrium. However, tornadoes predictably appear only when there are great differences of atmospheric (barometric) pressure. Indeed, tornadoes owe their very existence to the gradients of atmospheric pressure which they violently eliminate. By the time a storm ends, it has taken with it the pressure differences that were a condition of its existence.

Nature herself, without any help from a creative architect, fashions complex systems to break down pressure and temperature gradients. Real, three-dimensional complexity is not engineered, a byproduct of the algorithms of clever computer programmers. Real, three-dimensional complexity appears spontaneously—the natural result of ever-flowing energy finding shortcuts to reduce gradients in nature.

This is why life exists. Above and beyond the details of its genetics or its fascinating history, life exists as a thermodynamic means to degrade the solar gradient. Unlike the storm system, however, which reduces pressure differences in the atmosphere in a few hours, living matter has been reducing the Sun's gradient for some four billion years. Of course the gradient

PLATE 5
Tornado. May 13, 1980.
[*Howard Bluestein/Photo Researchers.*]

living matter breaks down is much larger than a summer thunder storm. Just as a spinning hurricane exists to reduce a pressure gradient, or the water-cycling whirlpool in your bathtub exists to reduce a gravitational gradient, so the recycling carbon chemistry of life exists to reduce a solar gradient. The gradient or difference in question is the difference between the hot, nuclear Sun and cool, cold space.

Life feeds on and accelerates the breakdown of this naturally occurring difference. Indeed, it has even been speculated that time's asymmetry—the apparent openness of the future, contrasted with the seeming closedness of the past—is itself an artifact of our dependence as life forms upon a thermodynamic gradient. Exactly how this trick is done may be beyond us. But Einstein's relativity theory tells us that true time is four-dimensional; it is symmetrical, with no privileged direction forward in time. Somehow, our parochial thermodynamic situation—living on the edge of a gradient, being part of this matter-cycling biosphere, this slow-acting, sunlight-storing storm—may be responsible for life's greatest misperception—the asymmetry of time.

At the same time, our sexual evolution is only coherent as a story, a story that begins long before genital-sniffing mammals or even any animal at all. It is this story that we tell here, beginning with the Big Bang and ending with the little bangs of our sexual descendants. Although there are local fluctuations, cells, organisms, and the biosphere as a whole all tend to break down the solar gradient. This, it appears to us, is the purpose of life in the same, no-frills way that the purpose of a tornado is to reduce a pressure gradient.

The long history of religious thinking in the West has led to a backlash against purpose-oriented (teleological) thinking in science. Organisms were once considered to be the way they were because God made them that way to fulfill a divine purpose. The triumph of modern, evolutionary biology, however, with its emphasis on accidents, contingency and adaptation, has made all talk of purpose highly suspect as religious atavism. Yet, in the most basic, physical way, organisms do display purpose. This is not divine purpose, of course, nor is it connected in any way with the Judeo-Christian god. Organisms act out a directionality, a goal-directedness in time that is intrinsic to their self-sustaining nature.

The "purpose" of autopoiesis, the self-maintenance we recognize as the simplest cellular forms of life, is to incorporate food into bodies using available energy. The first forms of life to evolve are thought to have been fermenting bacteria which metabolized sugars and other compounds forged naturally by the Sun's rays. The evolution from fermenters to photosynthetic bacteria, able to tap directly into the solar gradient in their matter-cycling activities, greatly expanded life's potential for breaking down and transforming stellar energy. Today, all growth and reproduction on the surface of this planet—the fantastic pattern we call life—is underlain by the Second Law and its tendency to reduce physical differences by any and all means available to it. Just as pressure differences don't always lead to tornadoes, so temperature differences do not always lead to life. But the purpose of both, in the most basic, natural sense, is to reduce differences.

THE NATURE OF DESIRE | This discussion leads to speculation on the place of sex in this physical universe. In closing this chapter we open our book with the idea that nature does have and has had a purpose. Even before the evolution of the first life forms, some four billion years ago—nature had a kind of want or desire. In most likely unconscious fashion, nature, as attested to in all the manifestations of the Second Law, wants to come to its ends. Life's reproduction, as we have seen, by producing order, makes messes and produces disorder as heat and local entropy. Now, for us, human reproduction, for hundreds of millions of years, has only been possible by a gender-based mating, by a winning sperm that fertilizes a waiting ovum and the subsequent growth of the fertile egg by cell division into the adult animal. Living matter is a peculiar kind of gradient breaker, one that is able to continue indefinitely due to the information-bearing repetitive chemistry of cell-enclosed DNA.

Just as we are matter contemplating its own evolution, so perhaps we represent, as sexual beings, the cosmos becoming aware of its own tendency to create and destroy. Sex is the beginning and end of that metacycle of carbon chemistry we recognize as an "I"—our own individual identity despite the imperfect repetition of our form in the generational mirror of our similar-but-not-identical offspring. In experiencing sexual temptation or pleasure, we enact a cosmic breakdown more primordial than life itself, one mandated in the very meaning of the Second Law of Thermodynamics. We help the Sun to spend itself, enhancing the degradation of energy as heat into space—not just by sweating in the bedroom but by giving the universe a chance, as best it can, to repeat our form. Because of their ancient tie to reproduction, our erotic acts, conscious or unconscious, achieve nature's ends. For nature's unconscious "goal" is to achieve the steady state, the state of maximal or near-maximal disorder, characteristic of gradient breakdown. Furthermore, nature, as shown in her construction of tornadoes, Bénard cells and living matter, is not above spontaneously forming highly improbable, beautifully patterned, materially cycling and rhythmically dancing structures in order to arrive at the most probable, disordered end state. In other words, all of our human purposes and wants, from the passing desire for something sweet to the burning desire of our life's greatest love, seemingly reflect inanimate tendencies already implicit before life in the Second Law. Sexual reproducers, live beings making more of themselves, achieve not only biological ends but those of physics as well.

The universe, one might say, is "in heat." Out of equilibrium since at least the Big Bang, the cosmos will remain so unless it is truly an isolated system of the kind studied by classical thermodynamics, in which case it will inevitably approach, if never entirely reach, stasis. In sexual pleasure we become aware of our own somewhat peculiar and contradictory penchant to simultaneously reach and not reach an end, to both break down a gradient and preserve it for further delectation. The desire to both indulge and preserve, to both hoard and spend reflects living matter's dilemma insofar as it both requires a source of energy but will disappear if it completely depletes that source.

Far from equilibrium, living systems are not independent but exist only if sufficiently near necessarily dwindling sources of high-grade energy. When chilled to near absolute zero ($0°$ K) living forms, if they survive, are "undead." Give them food and energy, however, and they revive. If rewarmed and wetted, freeze-dried bacteria, spores, cysts and other dormant life forms resume their metabolism. Indeed, such cryogenic experiments suggest that the degree of vigor correlates to the rate of energy and matter flow combined with the ability to handle the flow. Perhaps here, in the cosmic situation of life as gradient breaker, we can glimpse the material basis for the psychological structure of desire and the tension between instant gratification and prudent reticence. The paradox of life, including sexual life, is that the very frustration of the desire to reach a final end helps in prolonging it.

two

HOT AND BOTHERED: SEXUAL BEGINNINGS

Since Eve first chose her hellfire spark,
Since 'twas the fashion to go naked
Since the old Anything was created...

—William Blake

TRANSGENICS: THE WORLD'S FIRST SEX | SEX IS VERY OLD. Although they do not need it to reproduce, bacteria, the planet's first life forms, indulge in it. Indeed, the biotechnology revolution exploits the tendency of bacteria to donate and receive each other's genes: genetic engineering is based on the ancient sexual propensities of bacteria. All bacteria, because their cells lack membrane-bounded nuclei, are classified as prokaryotes. [PLATE 6] The oldest fossils of life, and its oldest chemical traces, appear in the rock record almost immediately after Earth formed a solid crust some 3.85 billion years ago. These remnants of the most ancient life are fossils of prokaryotes. Sex in prokaryotes fundamentally differs from the reproductive sex of animals and plants. Truly transgenic, prokaryotic sex always involves the movement of genes from a donor source (bacterium, virus, chemical solution or other) to a live recipient bacterium. This genetic movement, present at the dawn of life, provided an important means of survival for all subsequent life.

Prokaryotic sex originated during a geological period of formative violence and intense solar radiation. [PLATE 7] During the planet's early days, life was put to many tests. Bacteria were not only submitted to intense UV irradiation, they were also bombarded by cosmic rays, particles and meteorites from space. Because of its hellish landscapes, geologists refer to the tumultuous earliest era of Earth as the Hadean Eon. In the subsequent eon, earliest life as bacteria spread across cratered terrain, muddy waters and gurgling volcanic pools. This eon, in which the prokaryotes evolved, is known as the Archean Eon.

Bacterial life, forced into many different habitats, invented all the major forms of cellular metabolism, of which plants and animals utilize only a small proportion. Some bacteria, for example, are even able to use metals, such as iron and manganese, as energy sources. Bacteria evolved these abilities, in part, by gene donation, the world's first sex and still the most important for global ecology. In bacterial gene transfer, a donor bacterium

---

PLATE 6
Bacterial (prokaryotic) cell just before division. The entire process of reproduction is shown in Plate 12.
[*Kathryn Delisle/José Conde*]

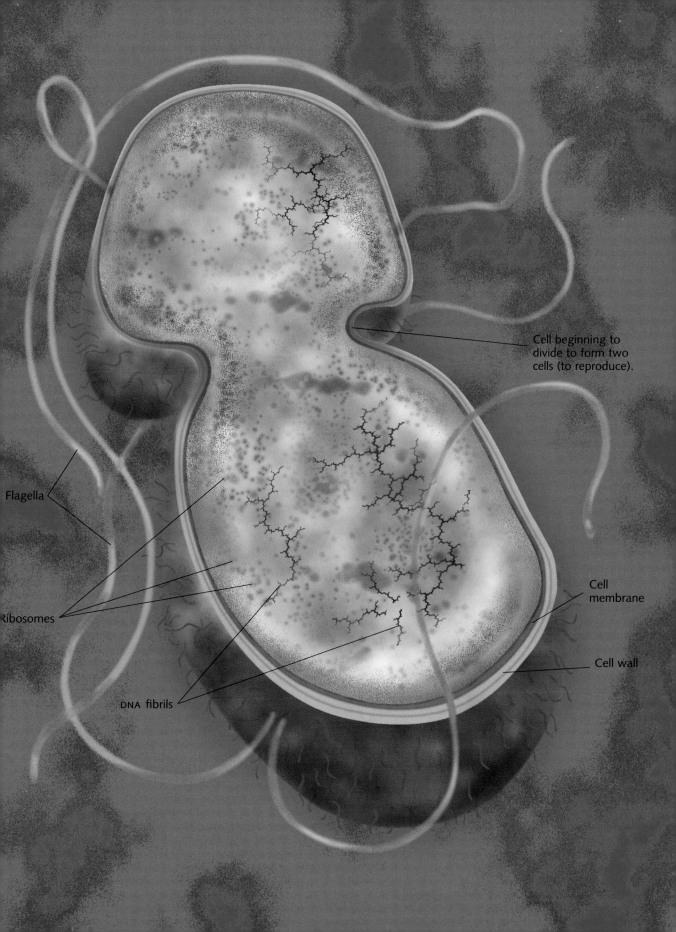

Cell beginning to
divide to form two
cells (to reproduce).

Flagella

Ribosomes

DNA fibrils

Cell
membrane

Cell wall

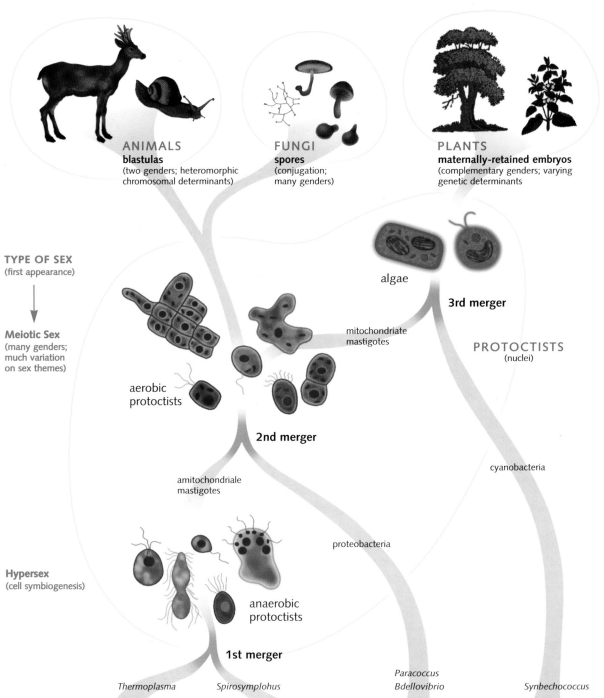

ANIMALS
blastulas
(two genders; heteromorphic
chromosomal determinants)

FUNGI
spores
(conjugation;
many genders)

PLANTS
maternally-retained embryos
(complementary genders; varying
genetic determinants)

Phanerozoic Eon (present)

TYPE OF SEX
(first appearance)

Meiotic Sex
(many genders;
much variation
on sex themes)

algae

3rd merger

mitochondriate
mastigotes

PROTOCTISTS
(nuclei)

aerobic
protoctists

2nd merger

Proteozoic Eon

amitochondriale
mastigotes

cyanobacteria

Hypersex
(cell symbiogenesis)

proteobacteria

anaerobic
protoctists

1st merger

*Paracoccus
Bdellovibrio*

Thermoplasma   *Spirosymplohus*   *Synbechococcus*

BACTERIA
(no nuclei)

Archean Eon

Transgenic
(bacterial)
Sex

fermenting
bacteria

swimming
bacteria

oxygen-breathing
bacteria

photosynthetic
cyanobacteria

ARCHAEBACTERIA   EUBACTERIA

52

passes one, a few genes or virtually its entire genetic endowment to its partner, and no reproduction or production of offspring is involved. Prokaryotes pass genes fluidly compared to sexually reproducing plants and animals. Plants and animals typically receive most of their new genes at the beginning of their life cycle in the act of their parents' fertilization, which leads to themselves as offspring.

Indeed, because the genes donated by one bacterium can be received by an extremely different bacterium—that is, because genes cross "species" barriers—University of Montreal bacteriologist Sorin Sonea and colleague Maurice Panisset argue that bacteria should not be classified as separate species.[1] If a species is defined, in the traditional manner, as a population of organisms that intrabreeds to form viable offspring, then bacteria do not qualify. They neither need to mate to reproduce nor are confined, when they do have sex, to pass genes to organisms within their own species. Different strains of *Bacillus anthracis*, *Bacillus cereus*, *Bacillus megaterium* and *Bacillus subtilis* can be very similar. Organisms interbreed; a population intrabreeds, but if two types of bacilli have more than 85% of their traits in common, bacteriologists treat them as members of the same species. Only 84%, however, and bacteriologists consider them members of different species. Counting traits and assigning species is pretty arbitrary when two strains of bacteria have from 80-98% of their traits in common. Even so, some bacteria, such as *Streptomyces griseus*, a multicellular bacterium that produces the antibiotic streptomycin, can receive genes from (mate with) *Escherichia coli*, a very different single-celled bacterium with nowhere near 85% of its traits in common. [**PLATE 8**]

The alien sexuality of bacterial "trans-species" gene donations is easily illustrated by imagining what would happen if it existed in mammals. If bacterial gene donation were a human capacity, a man with red hair and freckles might wake up, after a swim with a brunette and her dog, with brown hair and floppy ears. As often, scientific truth is at least as strange as

---

PLATE 7

Evolution of sex timescale. Symbiotic (hyper-sexual) mergers led from prokaryotes to nucleated cells (1st merger), oxygen breathing resulted from the 2nd (hypersexual) merger, and photosynthesis (algae) evolved from the 3rd merger. Fusion (meiotic) sex in ancestors led to animals, fungi and plants.

[*Kathryn Delisle/José Conde*]

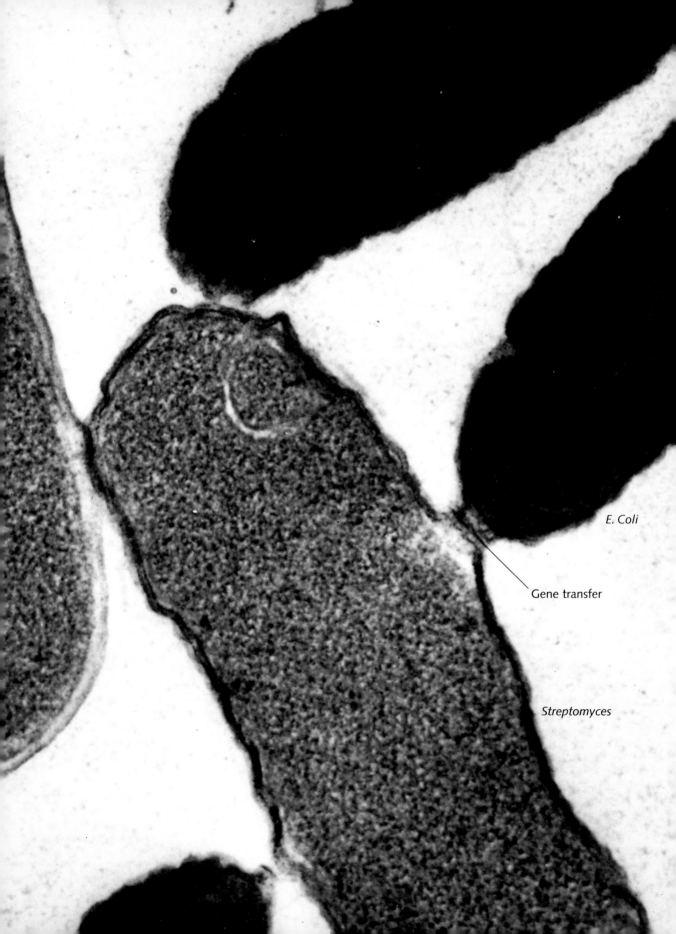

E. Coli

Gene transfer

Streptomyces

science fiction. Such monsters as the dog-human children of Moreau in H.G. Wells' *The Island of Doctor Moreau* pale next to the natural hybrids made by the rampant exchange of genes by bacteria in the real microworld. Faustian warnings notwithstanding, the "diabolical" biotechnology of "cross-species" gene mixing is older than species themselves.

Genetic engineers have borrowed, not invented, gene shuffling. The ability of one bacterial type to manufacture the proteins of another species allows us to use them to produce human insulin or pig hemoglobin and other substances normally made in nature only by those mammals. Genes for specific human proteins are combined with those extracted from bacteria and are put back into the bacteria to be expressed. This means that the hybrid gene directs synthesis of hybrid protein. The human protein's portion can then be cut out by specific enzymes. Fast-growing bacteria can be coaxed into making large quantities of human insulin or hemoglobin while they make their own proteins. We may frighten ourselves looking into the mirror of biology, but we should not mistake the image seen there for our own unrivaled prowess. Transgenics is far older, and much more than a merely human phenomenon.

Some types of bacteria are relatively stable, while others are veritable genetic charities, giving and receiving DNA to the point that it spells their own death. Indeed, when Oxford University zoologist Richard Dawkins, popularizer of the concept of the "selfish gene," was informed of the tendency among some bacteria to give away so many genes that they die, he was nonplused. But as Sonea and Panisset argue, bacteria are not really individuals so much as part of a single global superorganism, responding to changed environmental conditions not by speciating but by excreting and incorporating useful genes from their well-endowed neighbors and then rampantly multiplying.

Under death threats of too fierce heat, encroaching ice, a lack of certain nutrients, water loss and so on, access of bacteria to a common gene

PLATE 8
**Bacterial sex promiscuity I: E. coli transfers genes (arrow) to the filamentous Streptomyces, a very different kind of bacterium.
Photograph taken of thin slices of cells with an electron microscope. (See Plate 11.)**

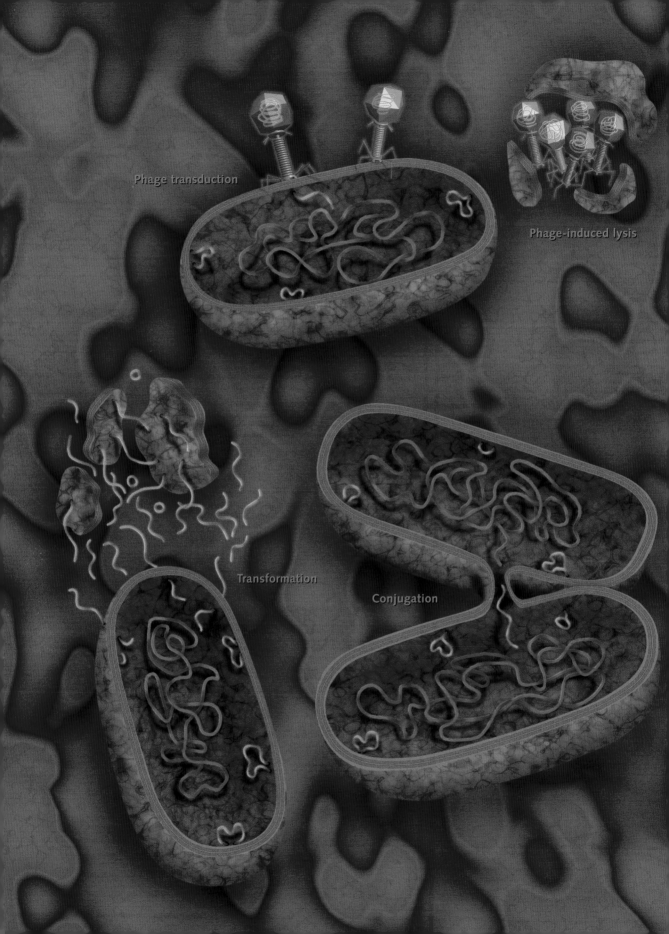
Phage transduction

Phage-induced lysis

Transformation

Conjugation

pool often saved early prokaryotes. Bacterial sex was a shortcut to survival. In evolution, the acquisition of a pre-evolved genetic complex—called a seme—can bypass the need to wait for fortuitous mutations. When Darwin proposed his theory of evolution by natural selection, he did not know the source of variation. Today, we know that the order of the chemical bases which make up DNA does spontaneously change or mutate. Usually these changes—like a misspelling or printer's error in a book—are deleterious. In an organism they can be harmful, even fatal. Occasionally, however, there will be a lucky mishap, a change that improves an organism's chances of survival.

Base pair changes and duplications, additions of extra DNA always have been a source of genetic variation. Now other sources exist. One is sex. In sexually reproducing animals and plants, however, the new genes acquired by sex are diluted when the species member breeds with a member of the opposite gender, necessarily mixing genes to produce a varied offspring. Bacteria, however, "breed true." In other words, because they reproduce without sexuality, if they pick up and incorporate new genes (by definition, sex) when they reproduce those new, sexually gained genes are passed on to their offspring without mixing.

A bacterium clones itself to reproduce. Metabolizing, it grows larger, copies its genes and the rest of the parts in its small body, and then "pinches off" (in a process biologists call binary fission) to make another, genetically identical cell. [PLATES 6 and 12] Like a cloned sheep, the new bacterium has only a single parent. Essentially, the new bacterium is the old bacterium, a three-dimensional carbon copy. Thus, if a bacterium gets lucky by acquiring a good batch of appropriately useful new genes, it can propagate those genes in its offspring without admixture. As long as the environment permits, it celibately clones any good new DNA it has received, and unlike usually useless random mutations, batches of genes taken wholesale from another organism have already proven their mettle. The dif-

PLATE 9
Bacterial sex happenings. Clockwise from top left to bottom left: Phage transduction; phage-induced lysis; conjugation; transformation.
[Kathryn Delisle/José Conde]

PARENTS

RECOMINANT

Donor Live Bacterium

In all cases the recombinant bacteria have DNA (genes) of their own and those from another source: the foreign genes come from a sexual act. Note these bacteria have not reproduced; they have just acquired new genetic material.

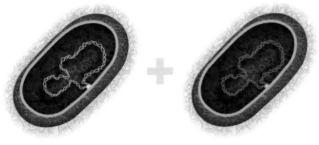

CONJUGATION

In **conjugation** the DNA of the donor (dark) replaces some of the DNA of the recipient live bacterium after cell-to-cell contact is made.

 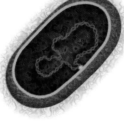 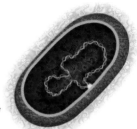

LYSOGENY

In **lysogeny** the donor parent is a bacteriophage (bacterial virus) shown monstrously large here. (See Plate 9 for more accurate relative sizes.) Bacterial DNA of the recipient parent incorporates a piece of DNA from the donor, a burst virus.

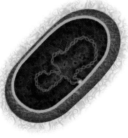 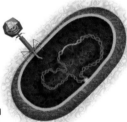 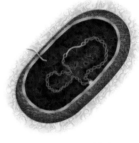

PHAGE TRANSDUCTION

In **bacteriophage transduction** intact bacterial viruses (bacteriophage) pass DNA to recipient bacterial cells. The viral DNA that incorporates into the bacterial DNA or chromoneme (large replicon often confusingly called bacterial chromosome.) The bacterial chromoneme is now described as "recombinant" since it contains viral DNA.

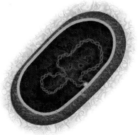 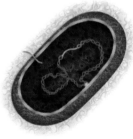

TRANSFORMATION

In **transformation** DNA released into the environment by dead bacteria is taken up and incorporated into the chromoneme of the recipient bacterium.

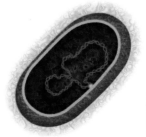 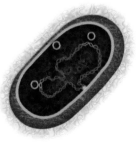

PLASMIDS

In **plasmid transfer** tiny circular DNA fragments (plasmids) are passed to recipient bacterial cells. The genes (i.e., for antiboiotic resistance can be immediately used by the recipient. The recipients thus can acquire heritable donor traits, such as antibiotic resistance, which are encoded by the plasmids.

ference is similar to that between a misprint, which almost always makes a text worse, and an appropriate quote, which serves a purpose. Genes containing information on how to metabolize potentially threatening wastes or toxins or on how to avoid dangerous compounds by swimming away from them were life savers. Like a Shakespeare quote in a modern text or a classical music riff in a rock and roll song, proven gene complexes—semes— proliferated in a new context. Bacterial populations allowed those that possessed them to dwell in otherwise inhospitable habitats.

So sex saved. Though tough and rapidly growing, any given prokaryote often lacked the resources needed to survive. As if the cosmic conditions in which early life evolved were not stressful enough, life repeatedly created further problems for itself whenever it rapidly grew. Wastes or toxic byproducts of metabolism are a thermodynamic inevitability. Rapid new growth, a sign of successful gradient breakdown, necessarily produces new chemicals which become a temptation to any kind of organism that evolves to make use of them. With prokaryotes came fresh answers to recurring environmental problems created by mutation and rapid growth. Bacterial sex allowed them to change, not only by accumulating mutations, but by receiving genes from their separately evolved, transgenic neighbors.

QUICK-CHANGE SEX ACTS | Bacteria indulge in several kinds of gene transfer, or sex: transduction, conjugation, and transfection. [**PLATES 9** and **10**] When bacteria lyse, they burst open: their cell walls and membranes explode and their genes are released. Fairly robust, naked DNA can survive suspension in water, heating, cooling, even freezing and other environmental insults. Prodigiously released by threatened bacteria, naked DNA can float in solution to reach live, still unthreatened cells. Such transduction involves the movement of bacterial genes, often in a ring, from the surrounding medium into a receptive bacterium. Bacterial conjugation involves cell-to-cell contact. A bridge or tube forms between cells by which

---

PLATE 10
Bacterial sex summarized. Sex requires two partners, the parents of the left-hand and center columns. The results of the various sorts of bacterial sex processes in related bacteria are shown at the right. These diagrams are cut-away views of bacterial sex acts.
[*Kathyrn Delisle/José Conde*]

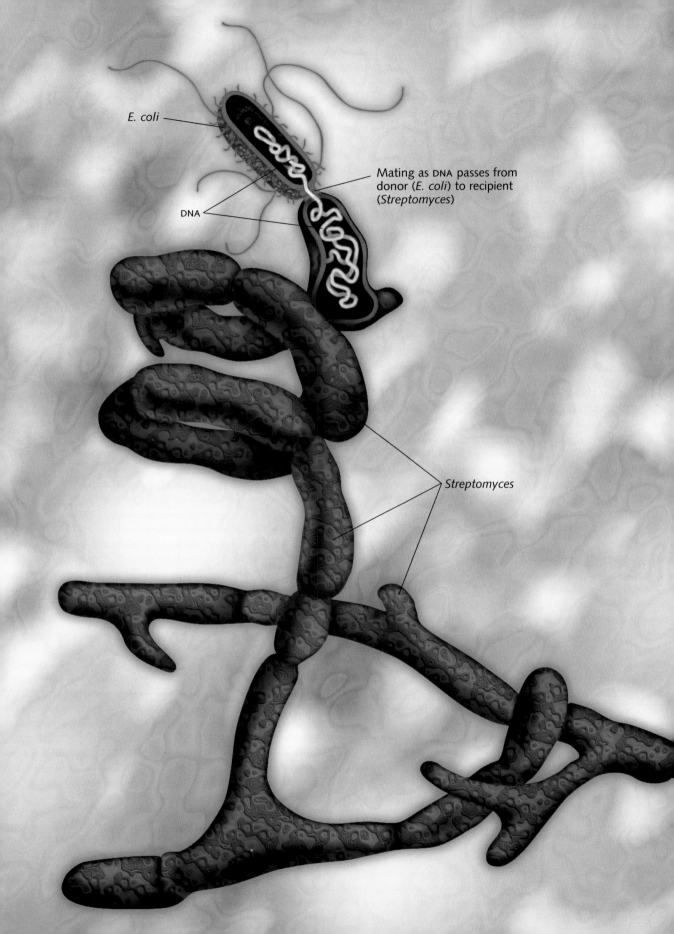

E. coli

DNA

Mating as DNA passes from donor (*E. coli*) to recipient (*Streptomyces*)

Streptomyces

anywhere from a few to many genes flow from one bacterium (the donor) to another (the recipient). [PLATES 8-11]

Genes can also be transferred between bacteria through a virally mediated process called transfection. Specific viruses which attack only bacteria bind to the outside of a bacterium and inject their genetic material into the cell, sabotaging the bacterial genetic system to make more viruses. The bacterial cell eventually breaks apart and new viruses, with combinations of their own and bacterial genes, are deployed to infect and inject this genetic material into other bacteria. As violently pathological and predatory as this sounds—and it is—it is sex. Sex requires the appearance of a new organism with genes from more than a single source—in this case from the original bacterium recipient and from the virus. Sex, acquiring new DNA, is clearly related to disease. Disease so often means acquisition of some other organism's DNA. The outcome varies—we call it disease if the new DNA endangers us. But often the new DNA helps the recipient bacteria live more effectively in a given time or place.

The ability of bacteria to sexually spread their genes was crucial in transforming the planet from a sterile, hostile place into one rich with a variety of abundant life. Rapid-fire ecological transformation ensued as different bacteria, with their great virtuosity of metabolic tricks, grew and changed. The single bacterial planetary metaspecies recycled compounds in many environments, including those that it itself created. Bacteria developed the ability to use oxygen and sunlight in photosynthesis and ammonia in nitrogen metabolism. They cornered countless other available resources or byproducts as energy sources to grow and thrive, including energy-rich gases (hydrogen, hydrogen sulfide and methane). The genes necessary to use these as energy and food sources were stored by some bacteria for eventual use by other bacteria. Bacterial sex encouraged global access to genes conferring a given metabolic ability, allowing virtuosities to spread worldwide.

PLATE 11
Bacterial sex promiscuity II. E. coli transfers genes to the filamentous Streptomyces, a very different kind of bacterium. Drawing reconstructed from photographs such as that in Plate 8.
[Kathyrn Delisle/José Conde]

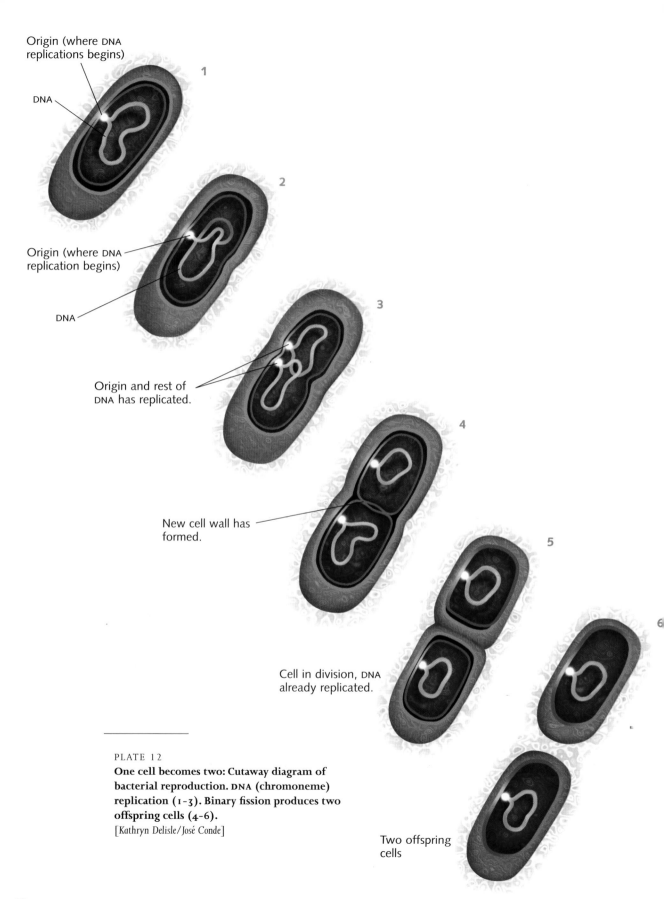

Origin (where DNA replications begins)

DNA

1

Origin (where DNA replication begins)

DNA

2

3

Origin and rest of DNA has replicated.

4

New cell wall has formed.

5

Cell in division, DNA already replicated.

6

PLATE 12
One cell becomes two: Cutaway diagram of bacterial reproduction. DNA (chromoneme) replication (1-3). Binary fission produces two offspring cells (4-6).
[*Kathryn Delisle/José Conde*]

Two offspring cells

Although plants, animals and fungi are relatively imposing simply because we are able to see them, this larger life is metabolically boring compared to the wild variety of metabolic tricks of gene-injecting bacteria. Plants, animals and fungi inherited their metabolic abilities (fermentation, oxygen-producing photosynthesis, and oxygen-based respiration) from bacteria. But bacteria retain a far greater metabolic repertoire than other life forms. Their ability to grow on yellow elemental sulfur gas or iron filings show our metabolism to be parochial. Only when we count industry and technology as human metabolism, do we even begin to measure up to the bacterial gift for environmental transformation. Yet even our much vaunted biotechnology is little more than kingdom pilfering, the borrowing, modification and specific application of a sexuality that has occurred naturally for thousands of millions of years among the gene-transferring bacteria.

Bacterial sex differs from ours relative to time. Bacterial sex, the first kind of sex on this planet, is speedy sex. Gene-excreting and gene-grabbing bacteria are still able to confer new traits immediately in the parent itself rather than through the parent, to the offspring. As stressed by neo-Darwinian theory, mutations, random changes in DNA, were important. Occasionally, mutated DNA allowed bacteria to access new food sources and survive at higher temperatures. But only with sex could a single bacterium access the genetic resources of the entire planet. With a little help from their friends, bacteria exchanged genes and were able to contingently respond to new environments. Sex gave bacteria a new lease on life. The ability to sexually transfer genes meant no bacterium was an island; each was, rather, a semi-independent cell in the biological equivalent of a global democracy, brain or supercomputer. This ancient environment-creating global system of bacterial sex continues to support our planetary ecosystem.

Rare mutant genes allowing important new metabolism do not remain rare for long because they are quickly copied and shared. Rapidly multiplying and rampantly transferring genes, bacteria conferred new skills upon their neighbors as well as on their offspring—a kind of genetic cultural transmission. Long before "the information superhighway," and long before human telecommunications or computer networks, bacteria formed an innovative, expanding planetary nexus of biochemical information. Indulging in non-reproductive sex, they broadcast useful genes across the planet.

SEX AND SOLAR RADIATION | How did bacterial sex evolve? The answer, reconstructed or inferred from today's laboratory observations, seems clear. Sex evolved from DNA repair. With no ozone layer on the early Earth, DNA was inevitably damaged by solar radiation. Hot and bothered by the Sun, whose gradient they reduced, photosynthetic and other bacterial forms of early life underwent gyrations in the nucleic acid control centers that are their genes. The Sun's rays ripped them apart and split them open. High-energy radiation mutated the cells, forcing them to extrude their DNA-rich fluids. Bacteria either died or evolved means of dealing with the incessant damage. But destruction of DNA by sunlight selected for genetic repair—fixing up and integrating one's own DNA. [**PLATE 13**] When, in desperation, the fixing up and integration of one's own DNA was possible only by integrating someone else's DNA, sex evolved.

Crisis seems to be a common denominator in the origin of sex in both bacteria and the large forms of life with chromosomes in their nuclei.[2] Life, an energetic process, depends upon, at a minimum, a common set of essential materials—water, carbon, nitrogen, sulfur and phosphorus compounds—and an incessant flow of energy. But life, whose ancient home is water, and which has encapsulated water in its move to land and air, is always in danger of drying out. The need for sunlight, at least by the vast majority of life dwelling at Earth's surface, has always been a Promethean game fraught with the danger of desiccation, over-radiation, and even

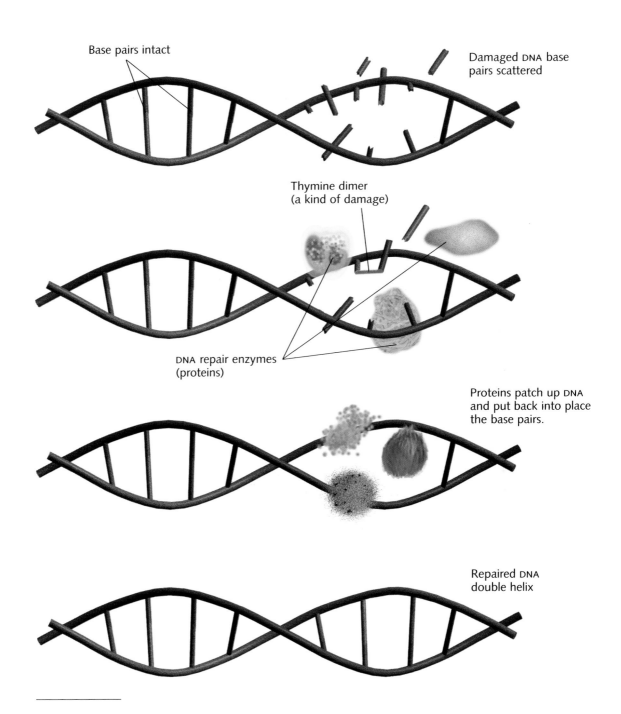

Base pairs intact

Damaged DNA base
pairs scattered

Thymine dimer
(a kind of damage)

DNA repair enzymes
(proteins)

Proteins patch up DNA
and put back into place
the base pairs.

Repaired DNA
double helix

PLATE 13

**DNA repair. A damaged double helix (top) is
acted on by two two kinds of proteins
(enzymes: upper and lower middle). The
result is repaired DNA (bottom). This is an
enlargement of part of the DNA shown as
squiggly lines in Plates 6, 9, 10, 11, and 12.**

[Kathyrn Delisle/José Conde]

spontaneous combustion. Whether transduction, transfection or conjugation, the bacterial sex process is nearly identical to the process of gene repair in bacteria bombarded by ultraviolet light. The enzymes used to patch up Sun-warped DNA became crucial for bacterial sex. These ancient DNA repair systems, from which bacterial sex seems to have evolved, are retained in all organisms to this day.[3]

That bacterial sex began over three billion years ago, when our atmosphere lacked free oxygen, we have little doubt. Without oxygen, no ozone layer existed in the atmosphere to protect genetic material from ultraviolet radiation. Data gathered by NASA's *Explorer 10* satellite from stars like our Sun suggest that the output of light energy during life's early evolution was so great that it is a wonder that any bacteria survived at all. Yet, under pressure of bombardment from both benign visible and perilous ultraviolet rays, life did originate—or at least persisted on Earth. Any bacterium that failed to repair its UV-damaged DNA perished.

The evolution of sex from DNA repair systems was probably serendipitous. Enzymes, such as ligase, which attaches or glues two DNA segments together, are crucial for emergency microsurgery in the wake of UV damage and necessarily evolved prior to DNA repair systems because of their vital role in DNA replication. Of course, DNA replication, necessary for growth and bacterial cell division, had been present in the earliest life on Earth.

DNA is the double-stranded molecule of genes, the genius molecule that, with lots of help from many protein enzymes, replicates itself. "Parental" DNA makes more of itself by first unraveling each strand. Then, with the aid of other enzymes (DNA-polymerases), the old strands direct the lengthening of new complementary "offspring" strands. Each of the "parental" strands now has a complement. If any of the double-stranded molecules becomes damaged, the information in the undamaged, functional strand may be used to replace the damage in the other. Like lines of type of the repeatable alphabet in a text file, DNA molecules carry around "back-up copies" of themselves intrinsic to their structure.

In the process of standard DNA repair, an organism uses a nuclease enzyme to cut out or excise the damaged portion. Then, nucleotides, the building blocks of DNA, are used to fill in the gaps. The undamaged strand serves as template to copy and produce a new, accurate double-stranded molecule. This splicing and copying from the "back-up" strand requires many repair enzymes, good timing and the loose DNA components called nucleotides. Once the anciently evolved DNA strand repair was refined to be an everyday occurrence in the bacterial genetic supersystem, it traveled widely across the globe. World-wide DNA repair systems opened the gateways to sex. In the simplest bacterial repair system, the gene or genes needed for repair came from one's own DNA. When outside or foreign DNA is integrated into recipients' damaged DNA, sex occurs. Offspring bacteria may have thrived with DNA from two parents, rather than one. Patching DNA—gene splicing—must have occurred often on an early, unprotected Earth.

When both strands of the single DNA molecule suffered damage at some site, in other words, when no back-up copy was retained because both strands were damaged and sequence information was irretrievably lost, cells died. No doubt many, if not most, cells died. But, early in planetary life, one bacterium used its repair enzymes to integrate foreign DNA from a mate or from the water. Ultimately, the DNA sequence from the other bacterium instead of its own was used: sex, as recombination, had begun. Like good news, the genes for these DNA repair enzymes traveled fast. The earliest integration of foreign DNA into one's own DNA was the first sex act. Sex, then, as a genetic infusion, permitted survival in a chemically chaotic and irradiated world. Such successful bacterial DNA transfer rescued bacteria, preparing them to repeat and refine their salutary sex acts.

Ultraviolet light still invokes quite a drama in bacteria today. Some bac-

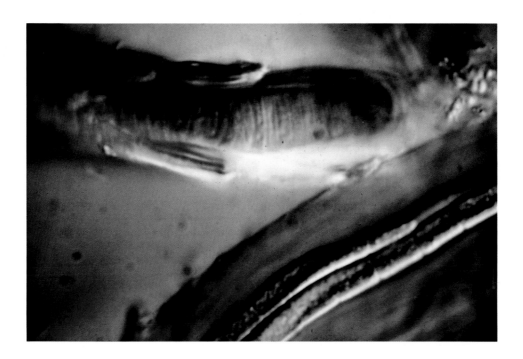

PLATE 14
Cyanobacterial "hats" shield organisms from
direct sunlight. Both *Lyngbya aestuari* (right
filament) and the coccoid cyanobacterium at
lower left are surrounded by such hat-like
protective sheaths.
[*Stjepko Golubic*]

PLATE 15
Sunglasses are sheaths that absorb excess solar
radiation by being dark colored. Top *Lyngbya*;
bottom right *Schizothrix splendida* with its thick
sheath.
[*Stjepko Golubic*]

teria avoid the blasting punishment: they possess the bacterial equivalent of "hats" or "sunglasses." [**PLATES 14** and **15**] Other UV-hit bacteria, resisting potential DNA damage, immediately stop growing, and release a suite of enzymes and naked snippets of DNA into their watery surroundings. They quickly make error-ridden copies of their long, UV-damaged DNA molecules. This prompt biological reaction to damage is called the SOS response. The SOS response, which also is brought on by environmental toxins, is an ancient means of ensuring the survival of at least a few bacterial descendants in the face of environmental crisis. Significantly, when bacteria lose the ability to repair UV damage, they invariably lose all sexual ability as well.

Escherichia coli, the human intestinal bacterium, repairs UV-damaged DNA extremely efficiently. But one strain of E. *coli*, called *rec minus*, cannot do so. This now-sexless mutant is also hundreds of times more sensitive to death by ultraviolet radiation than its sexually competent relatives. In all gene-donating and receiving bacterial sex, the DNA of one bacterium passes in one direction. The recipient takes up donor DNA, integrating donor genes into its own linear order. The formation of recombinant DNA requires at least one live bacterium and another source—dead or alive—of DNA. For the new, recombinant bacterium to survive at the end of this sex act, it must possess a complete set—one copy of each essential gene. The genetic recombination that enriches biotechnology entrepreneurs today started as a survival technique: first as DNA repair, then as naturally transgenic bacterial sex.

SEX BEFORE LIFE? | HOW OLD IS SEX? Evidence indicates that bacteria emerged soon after Earth formed a solid crust. But they did not evolve from nothing. Complex carbon chemicals, forged by the Sun and other energy sources, grew within dissipative thermodynamic structures.[4] There were complex, organized systems before life. Some scientists believe that there may have been the molecular equivalent to sex in this pre-biotic realm.

Fragments of DNA found their way in and out of protocells, in which case there is, in principle, no reason why sex, defined as genetic recombination, might not have existed even before life.

Sex is genetic recombination, the formation of a new being recombining genes from at least two "parental" sources such as donor and recipient bacteria. The proteins which make up all living bodies are not made directly by DNA but are assembled via RNA, the "messenger" and "translator" molecule that uses DNA information to arrange amino acids into proteins. In this sense RNA is a "sexy" molecule in that it can rearrange itself. Unlike DNA, RNA both replicates and directs the making of proteins. RNA's activity as an enzyme that chops and reforms itself is itself a strange kind of genetic recombination. Some see this to be sex at the molecular level. DNA, by contrast, must make RNA as its intermediate in coding for proteins. When the helical strands of DNA open, a portion of one of the DNA strands is "copied" onto messenger RNA. Transfer RNA molecules then use the messenger RNA as a template and match up the corresponding amino acid dictated by the code to form a protein. Whereas DNA is dependent upon RNA and enzymes in order to replicate, RNA can, in principle, make proteins without any DNA.

Some scientists postulate that, before the evolution of life, Earth was an "RNA world" of genetic fragments sloppily recombining and producing proteins which eventually evolved into life. If so, RNA may have been the precursor to life, a sloppy, on-the-way-to-coming-alive genetic information system.

Early Earth might have been a scene of rampant co-mingling among naked genes although it seems to us more likely that membrane-bounded protocells harbored the molecular antics of the ancient RNA strands. But ancestral RNA sex in any case—in the form of recombination—probably did precede DNA cells and help lead to the first bacterial forms. Sex—or at least its Sun-charged necrophiliac RNA-based precursors—could have come before life itself.

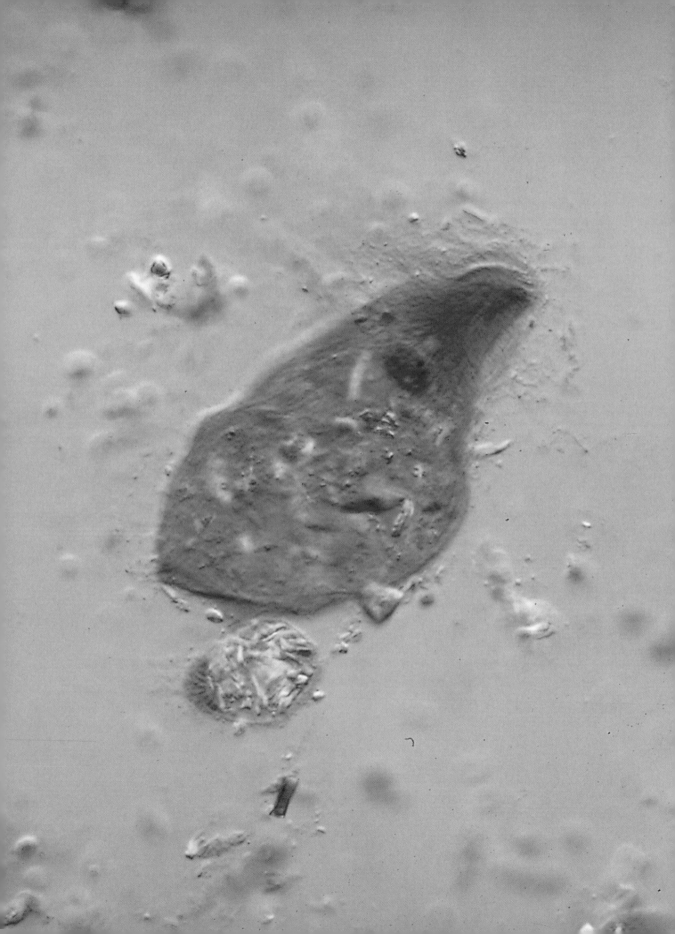

DANGEROUS LIAISONS: HYPERSEX | ALL FAMILIAR LIFE FORMS— large visible organisms such as plants, animals, mushrooms and lichens— come from "hypersex." We all know that, in sexual species, mates merge temporarily to produce new offspring with some genes from each sex partner. We are far less aware that many times in the past very different partners have merged permanently to produce new "offspring" with genes from each species. Instead of producing a new member of a preexisting species as in ordinary animal and plant sex, dangerous interspecies liaisons may produce entirely new species. Symbiosis, widely found in nature, occurs when two or more members of different species live together throughout most of their lives. Like sex, symbiosis brings partners together. But it can have a bestial element: it does not respect species boundaries. Analogous to the sex of bacteria, where, for example, *Streptomyces* bacteria may enjoy a sexual encounter with a very different cell, what we call hypersex—by definition—occurs between members of distinct life forms. Hypersex, defined here as permanent merging through symbiosis to make organisms with genes from more than a single source, has yet to gain full recognition in the traditional neo-Darwinian canon.[5] Nonetheless, it is a major source of evolutionary innovation.

The most striking examples of hypersex occurred in bacteria. One bacterium, entering another, grows and reproduces inside—forever. Permanent unions among originally separate bacteria led to new life forms, including, after hundreds of millions of years of evolution, all humans. Bacterial unions are the foundation of each animal cell in your body and each of the cells of plants. Your constituent cells, in other words, are hypersex hybrids. Except for a few tiny beings still inhabiting oxygen-depleted waters (like *Trichonympha* in **PLATE 16**), nearly all organisms composed of cells with nuclei—and these include protoctists (like ciliates and brown seaweed), fungi (like yeasts and mushrooms), plants (like ferns and wheat) and ani-

---

PLATE 16
Genderless Trichonympha. In this example of protoctist sex, the entire cell (of the hypermastigote *Trichonympha*) develops a gender and mates. The female forms a ring of granules at her posterior and the attracted male penetrates her from behind. His undulipodia and other organelles disintegrate, leaving only the nucleus, which mates with the nucleus of the female as seen in Plate 21.

mals (like clams and humans)—possess oxygen-breathing organelles. Cells contain tiny organelles called mitochondria which produce energy for the cell by metabolizing oxygen. These microscopic mitochondria (the little dark bars in **PLATE 17**] were once free-living, oxygen-breathing bacteria. In the early days before any animals, plants or fungi had evolved, small predatory bacteria, adept at breathing oxygen, probably forced their way into larger fermenting cells (protoctists) with no such capability. With time the invading agents became mitochondria. They permanently "mated."

Some hypersex begins as infection. For example, in the case of mitochondria, small invading oxygen-respiring bacteria failed to completely destroy their fermenting "hosts." Instead, they multiplied within them and kept their hosts alive. Paradoxically, the best infectors were not the most deadly. The "perfect" pathogen kills everything, then dies itself, for lack of resources. Mitochondria, bacteria that became organelles, most likely at first did kill nearly everything, but imperfection was essential as with time the disease became the new, permanently infected body. All our cells are marked by an ancient microbial bondage, the discipline of former infectors now trapped and working within our cells. Plants also contain permanently mated bacteria, the green-bodied former bacteria called chloroplasts. Chloroplasts began as free-living, free-floating walled bacteria called cyanobacteria(the third merger of **PLATE 7**). Even the most virginal plant or animal houses a promiscuous past, a long record of hypersex—permanent bacterial matings—deep within its cells.

Some hypersexual individuals approached each other in relative peace: one organism produced a waste product, such as oxygen or acetate, which the second organism could breathe or eat—and thus remove. In unity, oxygen-breathing, acetate-removing consortia stabilized. Bacteria communities became, through hypersex, new composite individuals. Other microbial associations, however, began contentiously. Aggressive, often fatally violent bacteria, such as, *Bdellovibrio*, *Daptobacter* and *Vampirococcus* are still on the attack.

PLATE 17

Protoctists from the hindgut of a termite. They ingest and digest wood. The black dots are the nuclei. In *Calonympha* there may be more than one per cell.

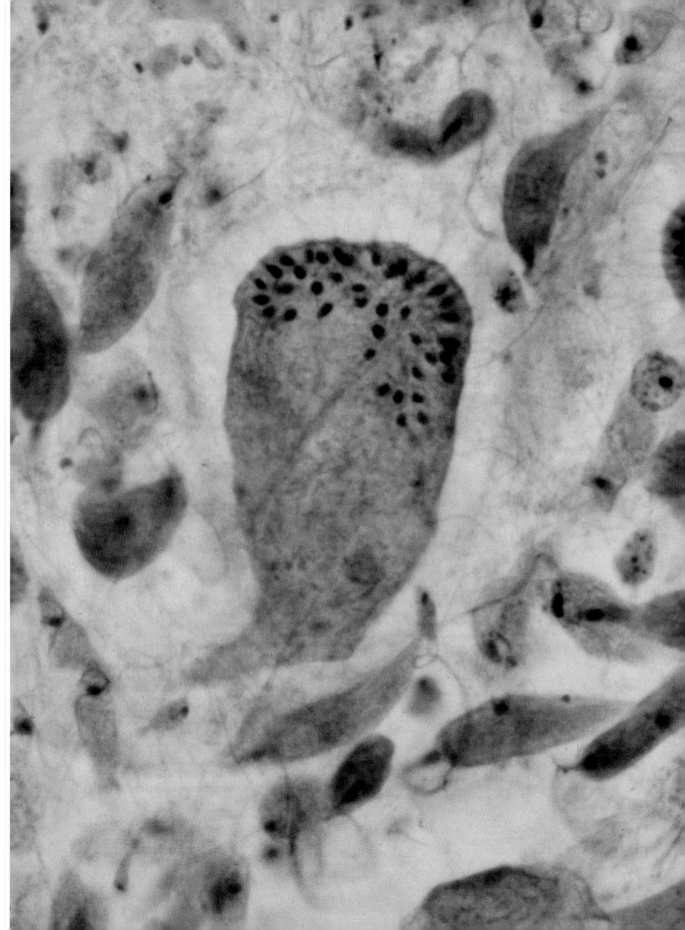

The protoctist eukaryotic offspring of bacterial hypersex, having more than a single type of bacterial ancestor, were structurally more complicated than their simpler unmerged single bacterial cell ancestors. A billion years of evolution of microbial communities led to weird and fancy hypersexual fusions. The protoctists, like *Staurojoenina* in **PLATE 20**, once ruled the world.

Once some of them began fusing and growing, they occasionally returned to their original unicell state. To persist, they had to do so each generation. The path opened for the evolution of sexually reproducing species. Fixed genders and sexually maintained species only began evolving about a billion years ago. Species such as dogs and cats are ultimately the result of the protoctists' invention of meiotic sex. With reproductive sex— cycles of doubling and singling in seasonally fusing protoctists—came the sexual species. In the next chapter, we will explore how these body-making, death-engendering cycles evolved.

ON THE WAY TO MEIOSIS | IN COMMON SPEECH sex usually refers to mammal's "genital friction." Perhaps because of the linguistic focusing on this simple three-letter word, sex sometimes seems as if it were a single process. Nothing could be further from the truth. Sex is multitudinous, complicated and confusing. It has an immense, deep history. At least three distinct kinds of sexual systems can be distinguished, all of which evolved in different organisms at different times and places. The first to evolve was the unidirectional type of bacterial sex that led to survival of a finely tuned global bacterial ecological network. Then, a highly specific form of symbiotic hypersex helped form our nucleated ancestors, the protoctists. Most recently in protoctist ancestors to fungi, plants and animals the most familiar form of sex evolved: meiotic and fertilization sex involving cell fusion. These were all necessary preludes to the growth of gendered bodies, such

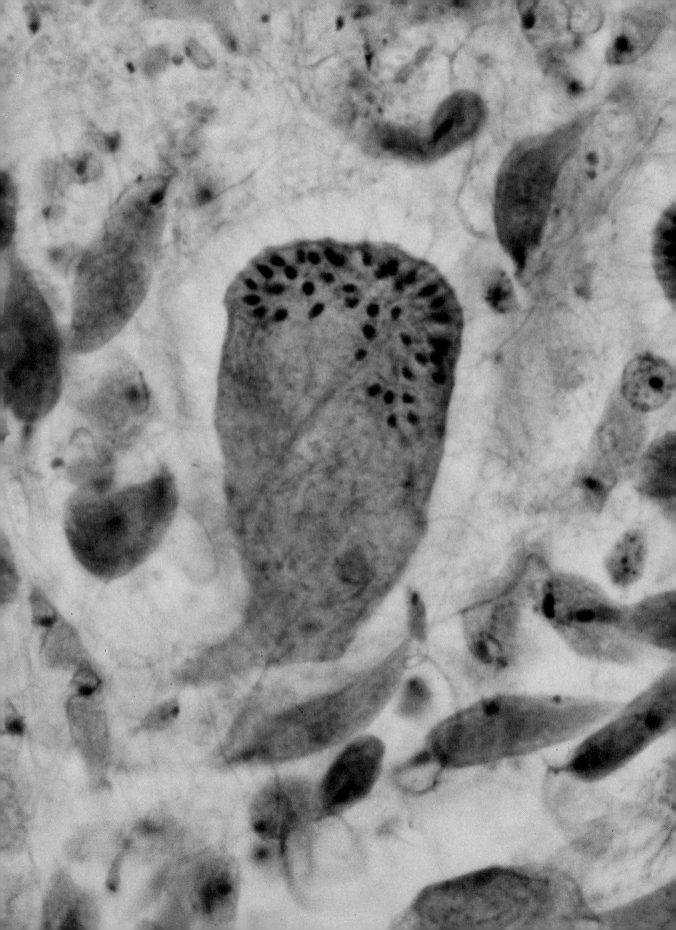

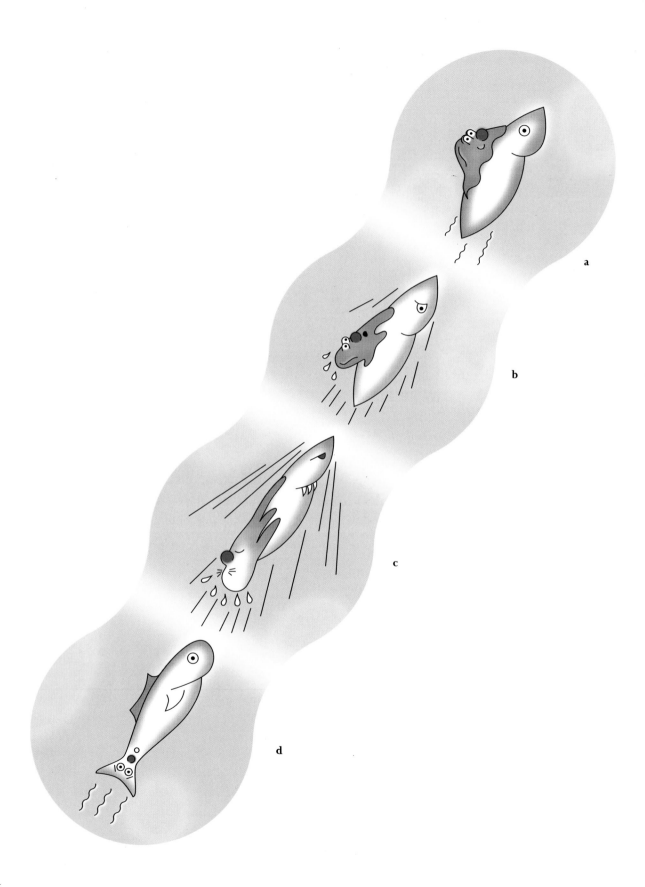

a

b

c

d

These bacteria invade and even kill their victims in their incessant search for sources of food and energy. Originally bacteria were often attracted to each other as a killer to its prey. But on a crowded planet, hostility is not always the best policy. Antagonistic organisms evolved metabolic truces that sometimes became genetic pacts. If you cannot beat them, goes the old saying, join them. Some bacteria, obeying this dictum, relinquished their autonomy and merged. More complex life forms were the result. The hypersexual move from predatory hostility to inseparable interdependence is comically illustrated in **PLATE 18**.

True hypersex, a permanent evolutionary embrace, almost always involves one partner who is a bacterium. The American anatomist Ivan Emmanuel Wallin (1883-1969) was an early experimenter in bacterial hypersex. Wallin was prescient but fallible. He was prescient because he hypothesized that our mitochondria originated from free-living oxygen breathing bacteria. He was fallible because he believed, incorrectly as it turned out, that he had removed mitochondria from animal cells and grown them independently. Mitochondria can be removed but will only survive for a few hours. They cannot be cultured—induced to reproduce when given food—on their own. Their bondage, and their discipline, is complete. Their individuality obliterated, they cannot live outside the cytoplasm of our cells.

Remnants of the permanent mating among certain bacteria we call "hypersex" persist in all cells of your body. Hypersexual miscegenation appears to underlie the origin of all familiar large organisms. Each of your cells is an amazing crossbreed, both more mixed up and more unified than anything found in a medieval bestiary.

HYPERSEX OFFSPRING AND SPECIES ORIGINS | AFTER BACTERIAL SEX came hypersex—permanent symbiotic mergers to form a new kind of cell—the kind with a nucleus. In hypersex, an entire bacterium enters the

---

PLATE 18
Symbiogenesis: Two different kinds of threatened beings merge to become one happy fish in four steps (a, b, c and d).
[*J. Steve Alexander*]

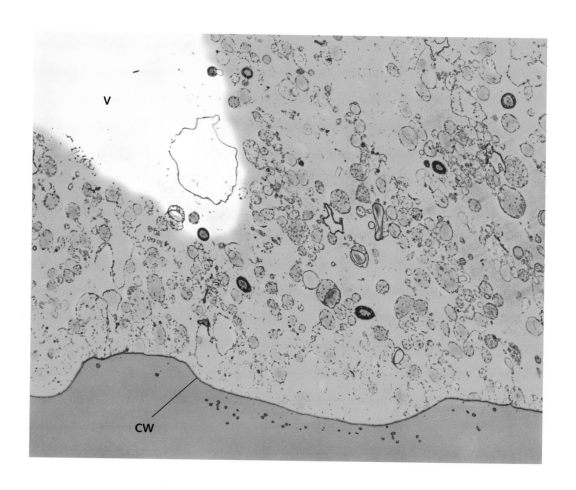

PLATE 19

Orange-colored *Beggiatoa* **from the Gulf of Mexico. This low magnification transmission electron micrograph shows peripheral cytoplasm with some endosymbionts of the larger bacterium visible toward the vacuolar (V) space. Some free-living bacteria are external to the cell wall (CW).**

body of another whole bacterium and the two types live together forever. The reproduction of hypersexual partners led to new units in evolution: cells with nuclei common to all nonbacterial life—from unicellular amebas to plants and animals with billions of such cells apiece. Some may protest that endosymbiosis is not sex. From an evolutionary point of view, however, it was even better than sex: such merged bacteria led not only to amebas, slime molds and paramecia, but eventually, after meiotic sex and its genders evolved, to all larger organisms, including us. Sexually coupling animals and insect-pollinated orchids are the beneficiaries of this hypersex ancestry. Indeed, animals, plants, fungi and protoctists have hypersexuality embedded in their cellular evolutionary history.

Nothing could be more intimate than sharing the closet space inside another organism's cell membrane. But this is what a few bacteria—similar in their lifestyles to our cellular ancestors—have been caught doing. [PLATE 19] Over evolutionary time, two or more bacteria inhabit the same living space, the same cytoplasm, and become inseparable. Eating and recycling each other's wastes, they indulge in transgenic sex as they pass their genes. Formerly independent bacteria merge entirely and permanently to become new, far more complex organisms. "Individuality" of large organisms is always intrinsically complex, the result of integration and loss of autonomy by more than one only remotely related ancestors.

The depths of hypersexual commitment reflect the natural history of bacteria. Normally, bacteria never fuse; they make contact briefly to send genes in one direction from one cell to another. But, in hypersex, they fuse forever. We animals evolved from the protoctists who are the hypersexual offspring—skipping maybe a billion generations—of the hypersexual protoctist "offspring" of permanently mated bacteria. The first hypersexual fusion of bacteria—between an obscure type of walless microbe belonging to the archaebacteria and a walled swimmer—led to the earliest of the nucleated cells—the first merger of PLATE 7.

The protoctist eukaryotic offspring of bacterial hypersex, having more than a single type of bacterial ancestor, were structurally more complicated than their simpler unmerged single bacterial cell ancestors. A billion years of evolution of microbial communities led to weird and fancy hypersexual fusions. The protoctists, like *Staurojoenina* in **PLATE 20**, once ruled the world.

Once some of them began fusing and growing, they occasionally returned to their original unicell state. To persist, they had to do so each generation. The path opened for the evolution of sexually reproducing species. Fixed genders and sexually maintained species only began evolving about a billion years ago. Species such as dogs and cats are ultimately the result of the protoctists' invention of meiotic sex. With reproductive sex—cycles of doubling and singling in seasonally fusing protoctists—came the sexual species. In the next chapter, we will explore how these body-making, death-engendering cycles evolved.

ON THE WAY TO MEIOSIS | IN COMMON SPEECH sex usually refers to mammal's "genital friction." Perhaps because of the linguistic focusing on this simple three-letter word, sex sometimes seems as if it were a single process. Nothing could be further from the truth. Sex is multitudinous, complicated and confusing. It has an immense, deep history. At least three distinct kinds of sexual systems can be distinguished, all of which evolved in different organisms at different times and places. The first to evolve was the unidirectional type of bacterial sex that led to survival of a finely tuned global bacterial ecological network. Then, a highly specific form of symbiotic hypersex helped form our nucleated ancestors, the protoctists. Most recently in protoctist ancestors to fungi, plants and animals the most familiar form of sex evolved: meiotic and fertilization sex involving cell fusion. These were all necessary preludes to the growth of gendered bodies, such

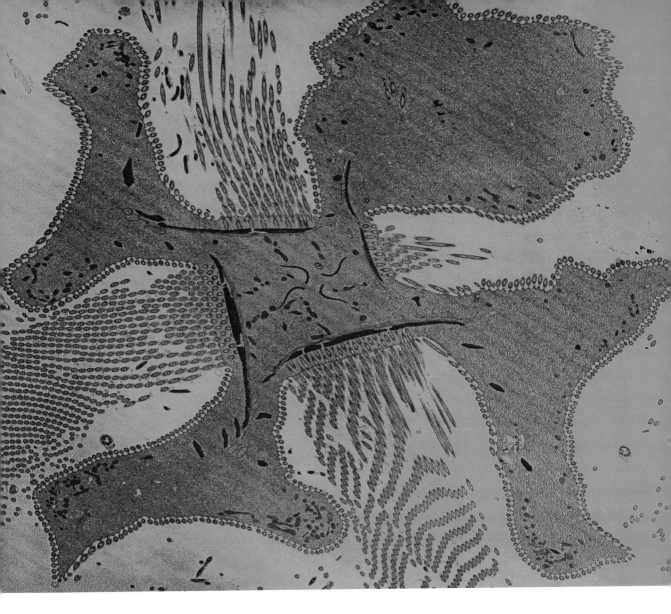

PLATE 20

Staurojoenina, wood-digesting four-lobed protist
from *Incisitermes minor*, Newbury Park, California.
A very thin section at the front end of the cell.
Each lobe is covered with aligned surface
bacteria and between the lobes are bundles of
undulipodia arranged in a cirlcle of 9 pairs
around a central pair—just like those of the
sperm of men and oviduct cilia of women.
[*Transmission electron micrograph by David Chase.*]

as ourselves. Without revealing the intimate details of our sexual ancestors, our own sexual urges, needs and proclivities remain obscure. Much is known about our sexual past but any scientist can only tell a tiny part of the story. We are trying to put together the whole tale from clues these scientists leave us. Our sexual history—in the broad sense—is bizarre.

Transgenic bacterial sex came first, followed by hypersex, the bacterial community mergers that instigated the new individuality of the nucleated cell. Each level of sexuality coexisted with and built upon the earlier modes. Life never forgets its ancestry. Transgenic bacterial sex, the sort that enriches biotechnology firms, probably evolved as a natural response to dangerous levels of mutating chemicals and ultraviolet radiation abundant on the hot, ozone-unprotected early Earth. Bacterial mergers, the specific symbiogenesis that we refer to as hypersex, blended distinct fermenting, swimming, oxygen-respiring and photosynthetic beings into permanent, if not perfect, unions.

The third type of sexual process, meiotic sex, evolved in those protoctist beings that had already evolved by bacterial hypersex. Meiotic sex produces propagules—the animal sperm and eggs, or the plant or fungal spores—by meiosis, the reduction-of-chromosome-number process. Meiosis is a kind of cell division. It reduces by half the number of chromosomes in an animal, fungal or plant cell. Meiotic sexuality—meiotic cell division to cut in half the number of chromosomes followed by fertilization to double it back—appears to be a balancing act of restoration. The advantages of leading a solitary lifestyle when life was easy was followed by the requirement to pool resources when times became tough. This dilemma and dichotomy—the single-double-single cycle of the chromosomes in cells—first evolved in the microbial world of protoctists. These protoctists behave today in ways that reveal their sexual histories, as we shall see. "You can't live with 'em, and you can't live without 'em," goes the old plaint of the lovelorn and heartbroken. The tension evinced by this statement may have roots, as we explain, in the somewhat schizophrenic compromise of the animal life cycle. We animals face an imperative, a balancing act between our existence as bodies with two sets of chromosomes that inevitably die, and sex cells made by these bodies with one set of chromosomes which enjoy the possibility of continued life in the next generation.

three

CANNIBALS AND OTHER VIRGINS: FUSION SEX

As no one has ever actually found perfect wholeness in another human being, no matter of what sex, the twin is the closest that one can ever come toward human wholeness with another; and—dare one invoke biology and the origin of our species?—back of us mammals doomed to die once we have procreated, there is always our sexless ancestor the amoebae, which never dies as it does not reproduce sexually but merely—serenely?—breaks in two and identically replicates.

—*Gore Vidal*

FUSING TO SURVIVE | Pardoned in advance for any unseemly consequences of your non-linear adventure, you step into a "time suit," select the time, and are transported to Earth as it existed 2,000 million years ago. You look at your monitor and see magnified, not only the left boot of your protective spacesuit-like gear, but some of your remotest sexual ancestors: two swimming, ameba-like microbes attempting to survive by fusing. By turns eluding and engulfing each other, they wrangle in the microbial equivalent of a wrestling match. [**PLATE 21**] Whether fighting or feeding, it is difficult to say, as each tries to swallow the other. Their wallowing seems as pleasurable as it is unsatisfying. You are fortunate to have landed at this particular place and time. For what you are now witnessing are the machinations that prefigured fertilization, the process by which an egg and a sperm unite to make one fertilized egg that grows to become an animal body.

Adjusting the knob of your portable microscope, you peer closely. At first you think that one swimmer has escaped the rather desperate embrace of the other, but now you realize that both are still there. They have fused. Or have they? The two, you observe, are almost one, but not quite. They have merged membranes, yet each still has its own nucleus. Together they have one body and two nuclei, and therefore, twice the number of chromosomes, twice the genes each had before the wrangling began. Once-separate selves, now doubled, they exist in awkward singularity, an almost-unified being. These widely separated nuclei move closer to each other. They seem attracted to each other. The nuclei, now very close together, fuse. Two organisms have become one; their two nuclei have become one larger swollen double nucleus. Now two nuclei float in one bloated body—one from each combatant.

You look up. The gas spectrometer on your time suit cuff indicates that the atmosphere, replete with hydrogen, methane and sulfide, lacks oxygen. Without your protective face mask, the poisonous gas cloud hanging over

PLATE 21

Early protoctist sex. Depicted diagrammatically are trichonymphids. Now gender has developed, and the male penetrates the female from behind. (See Plate 16).
[Kathryn Delisle/José Conde]

Undulipodia

Nucleus

Nucleus

Female

Fertilization ring
of female

Undulipodia

Male

the muddy scene would kill you. Toward the horizon, the Sun seems small and the planet strangely tropical, inhabited only by a voluptuous scum. No plant, animal or even fungus could survive this strangulating lack of oxygen in the atmosphere. But the struggling protoctist swimmers thrive. Some are fused in doubleness, others throb inside the membranes of their partner, as couples merge to stave off starvation. Some who have cannibalistically merged fight to the death. Other couples, even threesomes, get lucky. Fused indefinitely to their mates, they swim on in their doubled, or trebled, predicament. The luckiest couples have mitigated their doubleness. They have sleek bodies and only one—if swollen with twice the number of chromosomes—nucleus. If we were flip, we might call one half Adam, the other Eve. Or, to quote Bette Davis, "with separate bedrooms and separate bathrooms, I give them a fighting chance—[cackle]."

In the course of the wine-sipping philosophizing on the nature of love in Plato's *Symposium*, Aristophanes tells us of the primordial time when men and women—and men and men, and women and women—were single beings. Our doubled ancestors, he says, were shaped like balls with four arms and four legs which they could tuck in to roll faster than anyone could run. These whole ancestors were the androgynes, of which modern men and women are mere fractions. They were so powerful that Zeus, King of the Gods, punished them for their hubris. He split them in two. He tied the swollen umbilicus of each severed half-ancestor who, from that day forth, wandered Earth seeking its lost self.

DOUBLENESS | IF WE SUBSTITUTE the early protoctist for an eight-limbed roller, this tall story becomes not only provocative, but largely true. The hunger or thirst of our tiny ancestors occasionally drove them, in evolutionary rather than mythic time, to double and merge membranes rather than die. Although sexual passion does not arise, as Plato mused, from the endless search for our primordially amputated halves, it is due to a pri-

---

PLATE 22
Meiosis: The type of cell division in which the offspring halve the number of chromosomes of their parent.
[*Kathryn Delisle/José Conde*]

88

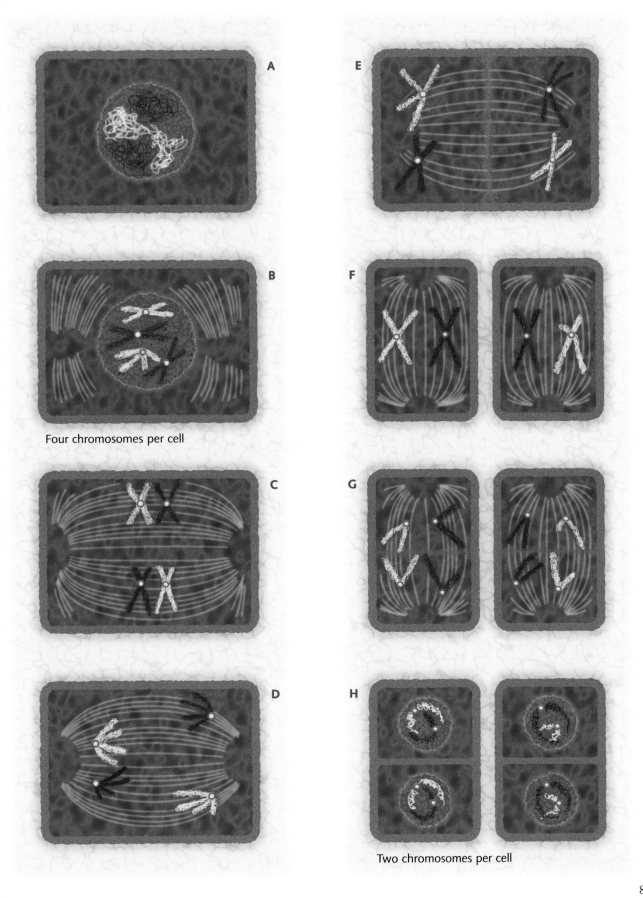

A

B

Four chromosomes per cell

C

D

E

F

G

H

Two chromosomes per cell

89

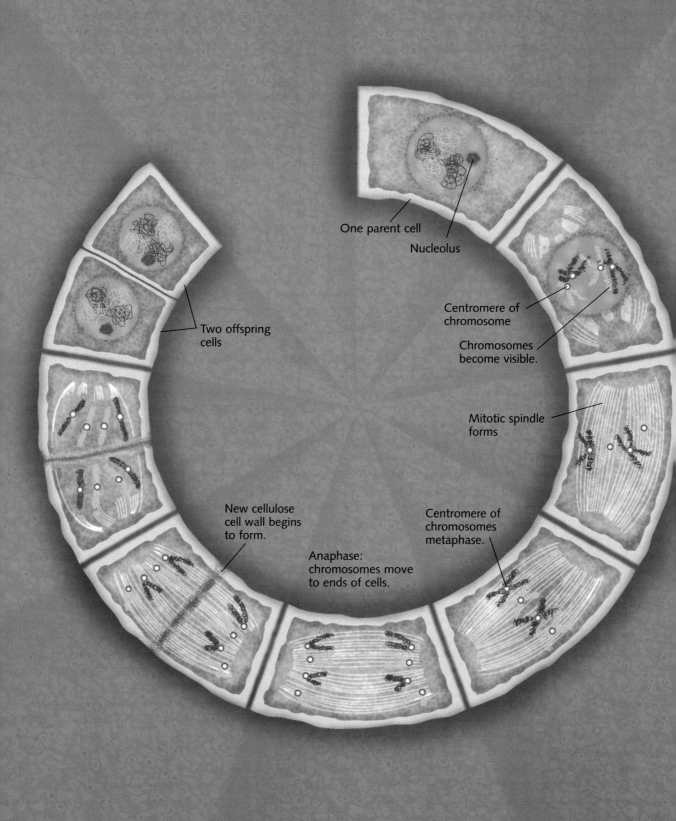

One parent cell

Nucleolus

Centromere of chromosome

Chromosomes become visible.

Mitotic spindle forms

Centromere of chromosomes metaphase.

Anaphase: chromosomes move to ends of cells.

New cellulose cell wall begins to form.

Two offspring cells

mordial doubleness. Each of our body cells is normally double or diploid. Like cells of any animal body, each of our cells possesses two sets of chromosomes. All cells, except for our egg and sperm cells, are diploid. Diploid refers to two sets of chromosomes in one nucleus, in people 46 chromosomes or 23 pairs. Haploid means possessing a single set of chromosomes, in people twenty-three. In animals, plants, fungi and many protoctists haploidy alternates with diploidy. As in all animals, our sperm and eggs, unlike our body cells, are haploid. Periodically they meet their counterpart and reestablish doubleness. In both plants and animals, the diploid nucleus, with its two sets of chromosomes, divides many times to form the embryo. The name for re-establishment of diploidy is familiar: fertilization. The name for periodic reversion to the haploid state is more technical: meiosis. [PLATE 22]

Meiosis (not to be confused with similar-sounding mitosis) is a cell division process that leads from cells with two sets of chromosomes (diploids) to those with only one (haploids). Meiosis is therefore often called "reduction division." Meiosis in men produces single-chromosome-set sperm from doubled body cells called spermatocytes. In women, meiosis produces single-chromosome-set eggs from diploid body cells called oocytes. Mitosis simply refers to equal cell division. A haploid cell with 23 chromosomes before mitosis becomes two cells each with 23 chromosomes after mitosis—the cell clones itself. Similarly, a diploid cell with 46 chromosomes after mitosis becomes two cells each with 46 chromosomes. [PLATE 23] In the meiotic process of reduction division, by contrast, one cell with 46 chromosomes forms at least two cells with only 23 chromosomes each.

Meiosis displays the several phases shown in PLATE 22. Animal reproduction requires fertilization to produce the sperm-egg fused pair that grows into the blastula, or animal embryo. In animals meiosis always makes sperm and eggs which eventually find each other, fuse and form fertilized eggs that become embryos. Haploidy, ending in fertilization, and diploidy,

PLATE 23
Mitosis: The type of cell division in which the offspring cells have the same number of chromosomes as their parent. Shown here in a plant cell.
[Kathryn Delisle/José Conde]

ending in meiosis, form the central cycle of the life history of animals. Although the details differ, meiosis followed by fertilization occurs in plants, most fungi and many protoctists. In animals, plants, most fungi and many protoctists mating is required at some time in the life of the individual to restore the diploidy destroyed by meiosis. Thus they (including us of course) are organisms that are said to undergo "meiotic sex."

After sexual union of egg and sperm in animals, the blastula's diploid cells grow as their constituent cells divide by mitosis. Nuclei multiply. Component cells elongate, bloat or otherwise change. The blastula, the multicellular embryo, develops a mouth, an anus, muscle and nerve tissue as it becomes a member of one of over 10 million recognizable animal species.

From protoctist alternation of singleness with doubleness, from cell fusions and relief from fusion by meiosis, there evolved large bodies. Plants also develop from embryos, multicellular bodies that live inside their mothers. These embryos enjoy their own developmental fates though they are not blastulas. Animal and plant bodies are always composed of many cells. To exist at all the cells of these bodies alternate between the unmated (haploid) and mated (diploid) condition. In those ancient, foul-smelling muds, away from what was, to them, toxic oxygen, doubled or diploid survivors—ancestors to animals and plants—would have been able to tolerate water and food scarcity more effectively than their single relatives. But, sooner or later, the double monsters would have had to recover their optimal form: sleek, effective, single beings with only one set of chromosomes.

Today, many protoctists still lack meiotic sex entirely. They reproduce directly from one to two cells by some successful variation of mitosis. Sleek and fast-moving, they are perfectly healthy with their single set of chromosomes. These haploid mitotic protoctists waste no time making a second, unnecessary, set of chromosomes. They never bother with genders, mating and sexual fusions. Even many algae, slime molds and fungi that can and do indulge in sex are truly healthy and grow perfectly well as haploids with their single set of chromosomes. As soon as they mate to survive a seasonal change or nutrient stress the fusion product forms a resistant structure, a progagule. Then, as soon as conditions warrant it, they revert—they relieve themselves of their "extra" chromosomes by meiosis.[1]

Unlike meiosis which halves the chromosome number, mitosis is the kind of cell division that maintains the *status quo*. The much more frequent mitotic divisions account for the growth of an organism when the cells stay together after the process. Mitosis leads to reproduction—one cell forms two or more cells—if the cell products of mitosis go their separate ways. Each chromosome replicates as its quantity of DNA and protein (the chemicals making up the chromosomes) doubles. The now-doubled chromosomes migrate to lie on the equator of the three-dimensional cell. Following this, each half-chromosome (chromatid) moves to an end of the cell (a pole). The two new offspring cells form as the now-doubled parent cell splits in two. The result of mitosis is two offspring cells, each with the same number of chromosomes as their parent cell.[2] This sexless mitosis, this reproduction of cells which themselves may be bodies, is performed by protoctists. Animal, plant and fungal cells also reproduce by mitosis as the organism grows. Of course, neither mitosis nor meiosis is present in archaebacteria or eubacteria whose cells lack chromosomes. (See note 2, Chapter 2.)

Intrinsically multicellular animals, plants and mushrooms grow through the mitotic reproduction of their constituent cells. These large organisms evolved from sexual ancestors who alternated fertilization with meiosis. But they probably did not all evolve from one single sexual ancestor. A preponderance of evidence (especially variation on the meiotic theme found in various protoctists) suggests that meiotic sexuality evolved at least several times, including among the distinct ancestors to plants, to animals and to fungi.[3] Ciliates, diatoms, green seaweeds, foraminifera, water molds

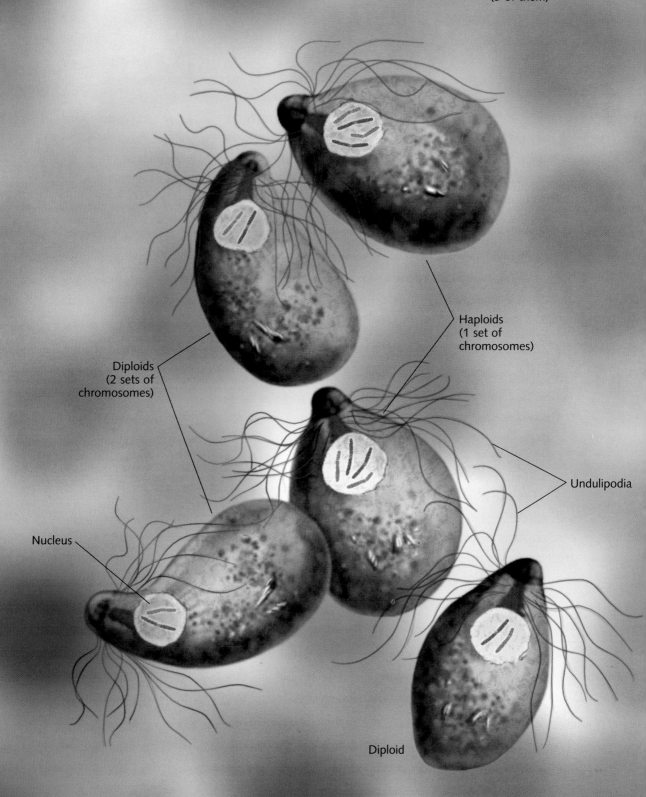

Haploids
(1 set of
chromosomes)

Diploids
(2 sets of
chromosomes)

Undulipodia

Nucleus

Diploid

and dinomastigotes presumably evolved meiosis in their own distinctive fashion. Of course, since all these protoctists grow and reproduce by mitosis, it is inferred that mitosis evolved prior to meiotic sex.

Mitotic cell division was, and still is, the eukaryotic cell's way of making more eukaryotic cells. In protoctists meiosis tends to be seasonal— a special kind of cell division which always must be balanced later by some kind of fertilizing fusion. Our ancestors were sexual protoctists, singles that when stressed merged, forming doubles. When the season changed they retreated to their former single state. The microscopically smaller protoctists, the protists, include beings such as amebas, euglenas and *Paramecium*. Amebas and euglenas still lack meiosis entirely, while paramecia have their own weird meiotic behavior. These living protoctists tell us about the kinds of ancestors in which meiotic-reproductive sex cycles first appeared. [PLATE 24]

Sexual experimentation in protoctists led to all compatible fertile matings in members of the same species on this planet. Species, as populations of consistent types of distinguishable organisms, had first evolved by the time bacterial hypersex produced the first protoctist some two billion years before. The earliest protoctists were nonsexual and reproduced by direct division, i.e., by mitosis. Some of their descendants were sexually indulgent organisms capable of fertile mating to produce doubled cells and subsequent relief of doubleness. By one billion years ago some protoctists seasonally practiced sex. Species then took on a new aspect. They stabilized by regular sexual fusion. The first sexual species had evolved. These were protoctists that seasonally ritualized meiotic separation (return to singleness) and fusion by fertilization (to remake doubleness). Such mating protoctists were (and still are) faced with a problem: to distinguish genders that were complimentary. Mates had to find each other to preclude the sloppy, even lethal cannibalism that had plagued their ancestors. Mates fused in each generation to recreate the only form of their living organization that could

---

PLATE 24
Protoctists, some haploid (one set of chromosomes) and some diploid (double monsters) looking for relief.
[Kathryn Delisle/José Conde]

survive the drought, the winter, the salt flat or whatever cyclical curse had led the threatened protoctist to merge in the first place. Although doubled, in crisis, they were better off as singles.

Early in the evolution of sexual species, the sex cells that fused were the bodies of the single-celled organisms themselves. Even today, sex cells of complementary genders often are indistinguishable from genderless, growing cells or the cells of the other sex. Protoctists, whose males (or other mates) look the same as females (or other mates), detect each other by very subtle cues. In the beginning, with cells as their only bodies, no specialized genitals or swimming propagules existed. Early mates looked just like each other. Over time, and separately in many lineages, equal single cells became distinct and unequal. Ultimately, anisogamy appeared in the form of small sperm and large eggs. With time different sorts of mating bodies evolved.

Several steps occurred on the way to developing the first meiotic sexual cycles. Likely the first was cannibalism. Even now, when smaller, foreign protoctist or bacterial cells attempt to enter and feed on the waste products or surface proteins of a large, single-celled protoctist, often they are neither digested nor rejected. Today, too, protoctists, when threatened, will engulf anything near them before they succumb. And under crowded conditions, many kinds of protoctists in frantic attempts to survive will even engulf each other. Such fusions—eaten embedded in eater—can, however, sometimes recover, especially if the two fused cells greatly resemble one another. The doubling up of chromosome number by sexual fusion requires an evolutionary explanation. Cannibalism, a common phenomenon in the protoctist world, provides one: phagocytosis—a fancy name for engulfment feeding—is frequent. Engulfment of single cells by neighboring relatives during hard times brought miserable company together. When both "partners" survived the result was an evolutionary step intermediate between feeding and the fertilizational doubling of meiotic sex.

In the 1960s, Harvard biologist Lemuel Roscoe Cleveland studied Trichonympha, Barbulonympha, and other protoctists. In one group, he discovered a type of cannibalism that could have been very similar to the first serendipitous origin of fertilization in our microbial ancestors. Cleveland, from 1934 until his death in 1969, studied wood-digesting swimming protoctists known as hypermastigotes. He carefully observed what he referred to as his "hairymen." He saw that, when starved, they were drawn to feed on their fellows. The attraction, however, was not sexual since these hypermastigotes routinely reproduce mitotically without gender or sex. The attraction was more the cannibalistic attraction of a Hannibal Lector in *Silence of the Lambs*. Desperate hypermastigotes came together in attempts to devour each other. Waving their multitudinous swimming appendages (undulipodia) as they swam in close to feed, they sometimes adhered. Once attached, one opened its membrane and fused with the other, as if pursuing its normal habit of taking up pieces of wood for food. Of course these beings, unlike Hannibal Lector, were not criminally insane. They did not attempt to kill each other for depraved sport. Without food or energy sources, the starving hypermastigotes were destined to die. Eating close relatives—cannibalism—sometimes saved them—a final, desperate meal, tiding them over until conditions improved. Whenever desiccating or frigid conditions reappeared, cannibalism increased its stock as a viable survival strategy.

But the significance for the history of sex—and the reason it is, in some sense, less cannibalistic than vampiric—occurred in the aftermath of the frantic action of the moribund hypermastigotes. Cleveland saw that, in some cases, one fused protoctist would not be fully absorbed by the other. Rather, sometimes the engulfed desperado continued to metabolize in an effort to survive inside the body of its abductor. The victim was weakened but not killed. The would-be "food" did not die. Without immune systems, microbes easily fuse. In some cases Cleveland observed a merging of the nuclear and cell membranes so that one fused cell with two sets of chromosomes existed where before there had been two separate living cells.

Noticing the fundamental similarity to mating—which also involves cell-to-cell attraction, cell membrane merger and fusion of nuclear membrane to make doubled (diploid) nuclei—Cleveland speculated that his laboratory "hairymen" had experienced the sort of event that first brought cells together in the origin of meiotic sex. Swallowing each other, the mastigotes merged nuclei and cell membranes. Attempting to feed, they fused instead. This, Cleveland theorized, was how fusion sex, the meiotic-fertilization cycle, began.

But other pieces to the puzzle lay unconnected. The abortive cannibalistic act of fusion may have doubled the number and made a diploid monster, but then how was the sleek single-set-of-chromosomes hypermastigote to recover its original haploid form? Continued indulgence in even more partial cannibalism—stimulated by the misfortune of thirst, cold or starvation—would increase the chromosome number and further swell the already swollen hypermastigote body. What was needed was recovery—reduction or diminution of chromosome numbers. To evolve into our mating ancestors, the miniature monsters would have to be relieved, as Cleveland saw, of their cumbersome burden, their diploidy.

He recognized that whether thirst, starvation, heat, cold, high salinity or other environmental insults had provoked fusion by partial cannibalism, relief of the partially cannibalized, vampiric state, was necessary. In the end, the relief that followed when fair weather and plenty again prevailed could only be reliably obtained by meiosis. But how did meiosis evolve? Cleveland devised a scheme for how meiosis might have evolved fairly simply from mitosis. The main thrust of the evolution of meiosis from mitosis entails timing changes, a temporary delay of chromosomes to replicate.

The doubled monsters could not persist indefinitely. When the chromosomes replicate, timing mishaps are normal. Any delays that led to a reduction of chromosome numbers would have been strongly and positively selected. Thus, after cannibalistic famines, when conditions of plenty

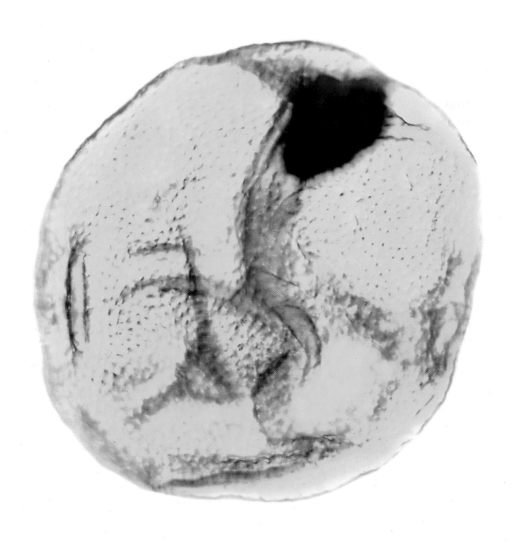

PLATE 25
Fossil, seen with a microscope, an "acritarch,"
a name which means that no one is sure what
it was. Most acritarchs were probably sexual
propagules of ancient protoctists.

returned, the ancestral state of one single set of chromosomes (the haploid state) would once again become optimal. Protoctists had flourished for millions of years as solitary swimmers, multicellular algal sunbathers and stringy water-mold feeders. They still do. Protoctist life had been honed in the haploid, single state. Doubling, while it saved lives during hard times, was an impediment during times of ready provisions. Today most mated protoctists make some kind of overwintering structure, a resistant propagule. [PLATE 25] In protoctists, mating to make a doubled being does not, as in animals, start the growth of an embryo. Rather, all it usually does is lead to a protected diploid structure able to withstand environmental threats. Physiologically, such propagules seem to represent a flashback to leaner times.

Mating thus led to fusion which, in turn, led to dormant survival. But when good times returned, the original protoctists with a single set of chromosomes—the haploids—were more adept. They reproduced rapidly. When summer returned, it was of great advantage to be single. Modern beings provide clues to the circumstances of early sexual fusion. Former mates, now fused, form cysts, zygospores, hystrichospheres, oocysts and many other sorts of propagules, depending upon the species. Such resistant structures offer protection, perhaps partly by radically lowering the energy and material flow-through rates. Microbial sex, in other words, induces the microbial equivalent of hibernation: given the seasonal disappearance of gradients necessary for life, the formation of sexual propagules tends to save lives.

Sexual animals, at root populations of mitotically reproducing cells, evolved from early protoctists that at first reproduced sexlessly. But early sex by fusion, relieved every season by meiosis, led to a survival strategy that worked and was consequently used again and again. Then came another innovation. Instead of forming dormant cysts by sexual fusion, our non-animal ancestors grew in the doubled state—they formed bodies, doubled

PLATE 26

Hystrichosphere. This fossil is like an acritarch, but openings on its surface permit its identification as a sexually produced cyst that was made by ancient dinomastigotes. Dinomastigote marine whirlers (dinoflagellates), are a kind of protoctist that left hundreds of fossils and many living descendants.

[Kathryn Delisle/José Conde]

"Decorations" on hystrichosphere,
perhaps to aid flotation (no one knows).

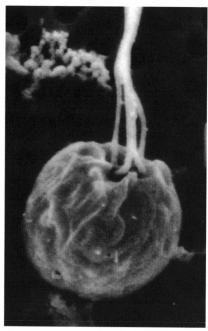

A

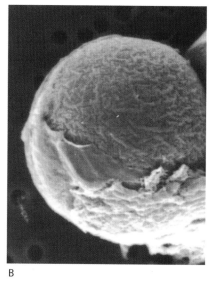

B

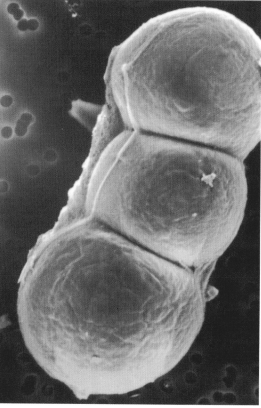

C

PLATE 27

Chlamydomonas zygospores.

A. After mating and fusion the *Chlamydomonas* doubled monster swims awkwardly with four undulipodia.

B. Swimming ceases and the fused, now diploid cell develops a hardened thick wall as seen here in this scanning electron micrograph of one zygospore.

C. An entire culture of cells were mated. Six mates formed the three clumped diploids seen here. They fused by twos, formed quadriundulipodiated swimmers (like the one in A above) and settled down to make these thick-walled zygospores. These zygospores are stable and successful in resisting unpleasant environmental conditions for many months, even years. As zygospores, *Chlamydomonas* survives the starvation of cold winters, salty water, dry hot summers or other environmental insults that would kill an unmated swimmer.

[*Charlene Forest*]

but no longer monstrous—that could escape the conditions that led to the misery of moribund merging in the first place.

SOUL SURVIVORS | THAT PROTOCTISTS ENGAGE in sex to withstand hard times makes perfect sense. Protoctists that fused under crisis conditions—conditions that killed their softer, unfused neighbors—survived to evolve the hard, desiccation- and cold-resistant thick-walled propagules shown in **PLATE 26**. Such crisis fusion was the precursor to sexual fusion. Even today, these cysts do not begin to grow until food and water reappear or conditions improve. Sexual fusion in protoctists does not result in reproduction. Rather in many green algae and dinomastigotes, for example, sex leads to these hard-walled structures with reduced metabolism. Fusion brings on the winter cysts that lay dormant until spring.

The most fascinating clue to the multiple origins of our kind of meiotic or reproductive sex comes from the fact that so many species of protoctists are actually induced by environmental stress to undergo sexual acts. For example, the common green pond alga, *Chlamydomonas*, some of whose red relatives inhabit snow, quickly reproduces by mitotic cell division. Usually, it has no need for sex. Deprive it of nitrogen, however, and it will seek out an identical-appearing member of its own species and will mate with it. The two fuse and, instead of each having two waving appendages called undulipodia, they form a swimming fused quadri-undulipodiated "monster." Nitrogen depletion induces sex, and sex creates an ogre which, in five days, becomes the black, hard-walled *Chlamydomonas* zygote of **PLATE 27**. This zygote can only be released from its sleeping spell by meiosis. Submerged in water containing nitrogen salts, the haploid meiotic bi-undulipodiated cell swimmers emerge and swim frenetically away.

Even in the laboratory the *Chlamydomonas* zygote must wait at least a week before any stimulus can induce it to begin to grow again. The cyst, with its hard exterior, resists deprivation like a hibernating bear or bat, like a walnut, tardigrade tun, or fungus spore. It buys time. In its state of suspended animation, metabolism lowers to match leaner available reserves: sex is economical. Zygote and other cyst formation is a cunning, if somewhat desperate strategy. Microbes form a cyst through mating, then relieve that doubleness inside the cyst through meiosis. *Chlamydomonas* and many other microbes routinely sacrifice their present growth prospects for the promise of future growth. Anticipation of coming environmental hardship evolved in protoctists long before animals or consciousness as we know it appeared on the evolutionary stage. Indeed, planning, touted as a consciousness-based, uniquely human characteristic, implicitly exists in microbe cyst formation. It is an evolved physiological response to seasonally fluctuating food and energy reserves.

All protoctists reproduce mitotically. They grow perfectly well without sex. Since no fertilization ever occurs, no gender differences are developed. To wait out tough times, others quickly mate and form dormant propagules full of future prospects. At the basic, eukaryotic level of the single protoctist cell, sexual mergers figure as a last resort, an emergency measure that greatly increases an "individual's" chance of survival. We use quotes here because such encystment, followed by fertilization, shows the strange interdependence of one with another. In the prehistory of sex, only those who coupled evolved.

BODY BUILDING | CYCLICITY WAS A CRUCIAL INGREDIENT in the story of the origin of reproductive sex. Today, animals, plants and fungi—and all sexual protoctists—are neither wholly single nor double. Rather, we are always both. As animals, we cycle between the doubled state of body cells and the single state of sex cells. The body cells of fungi are in the single haploid state. The cells of mushrooms in the northern cool region of the United States cyclically form the double state in late August or September. Meiosis immediately relieves this doubleness as single-set haploid spores are scattered to the winds in late autumn. Such cycling between singleness and doubleness (haploidy and diploidy) began with the establishment of meiotic sex. In all obligatory sexual species, it is necessary for cells to be sometimes haploid and diploid at other times. The selection pressures for each distinct state originally may have cropped up on a regular basis. Seasonal cold, predictable droughts and other cycles alternately provided and deprived microbes of a gradient upon which growth depended. Many lineages of successful protoctists discovered sex and found it beneficial. These protoctists revert back and forth between an ancestral quick-growing haploid and a hardy diploid state; they toggle between thriving and surviving.

The first sexual reproducers likely alternated between a harsh (winter or dry) season in which they were fused and a summer fast-cruising in which they were single. Growing plant and animal bodies are intrinsically more complex: they start where fused protoctists leave off, becoming embryos. The brilliant early plant and animal body strategy was to grow in the interim, fused state. Such growth as multicellular diploid embryos opened up whole new ecological possibilities before the necessary return to the ancestral spore or sperm haploid.[4] Today, what we think of as a plant or an animal body is always a mass of mitotically reproducing cells.

A strange, but illustrative, example of a many-celled body cycle involves protoctists called cellular slime molds. These wet, creeping, mucoid masses display seasonal fusion followed by fierce individuality. Their lifestyles provide us with insight into the "missing links" between starved individual microbe cells and large feeding and growing cell masses theoretically similar to the ancient societies which evolved into the first bodies.

Until recently lumped with plants or fungi, the slime molds, both acellular (myxomycote) and cellular (like *Dictyostelium* or *Acrasia*) types, are now recognized as unique protoctists.[5] Unlike fungi, they lack hyphal threads and do not have cell walls composed of the organic material chitin. Unlike

PLATE 28
Sex in many protoctists leads to stalks and spheres that house propagules. These were made by swimming and crawling protists called myxomycetes that mated and grew.

plants and animals, they entirely lack embryos. Slime molds are superb examples of why Protoctista deserve recognition as a separate and unique group. Myxomycotes such as *Echinostelium*, *Fuligo*, *Lycogala* and *Stemonitis* today are placed by most taxonomists into their own phylum in the protoctist kingdom. [PLATE 28] Arguments of whether these moving heterotrophs are animals or plants have been resolved: they are neither. Their peculiar sexual fusions and alternation between multicellular and unicellular bodies illustrate weird side branches in the evolutionary tree of sex.

Myxomycote slime molds move as feeding cell colonies that alternate between single and multicellular states. Found on fallen or decaying wood on the forest floor, on wood chips of garden paths or on compost heaps, they are microscopic in their single cell stage but easily detectable in their stalked, dark, spore-bearing stage. Each delicate spore stalk, some more than a centimeter in height, persists for at least a few days.

Myxomycotes alternate between a reproductive feeding stage, in which nuclear division outpaces cell division to form multinucleate amoebaelike creatures, and a dormant stage. The many-nuclei slime creature, called a plasmodium, expands and contracts, throbbing slowly in search of victims. Bacteria and tiny protoctists, trapped by the throbbing plasmodium, are eaten alive. [PLATE 29] When full and satisfied, the plasmodium settles and a reproductive phase of sedentary spore stalks ensues. Nuclei divide to form walled spores. This alternation between the slimy plasmodial stage made of multinucleate, flowing non-cellular cytoplasmic masses, and the sedentary, spore-bearing stage made of single-nucleated cells, is typical of these colorful plasmodial slime mold organisms. [PLATE 30]

We call the change from caterpillar to butterfly "a metamorphosis," while we refer to the change when two adult butterflies die after mating as death. But, from a deeper perspective, nothing has died. No egg-caterpillar-butterfly metamorphosing life has been lost. The butterfly's "death" is a natural phase, the next developmental step in a sexually mediated, cyclical

PLATE 29
Another myxomycete, the yellow sporophores
of *Fuligo septica* and *Dictydium cancellatum*
sporocarps), provide evidence that sex
occurred in the soil.

PLATE 30
Physarum (yellow), a myxomycote plasmodium
feeding on bracket mushrooms.

form of cell organization. In myxomycotes, whose bodies are less distinct than those of insects, the persistence through change of many life forms is clear. When these slime molds' black spore wall splits, their sexual life cycles "begin." The spore releases a free-swimming ameba-like amoebo-mastigote. Its nuclei divide and divide again to form the oozing, membrane-bounded protoplasm; the life cycle continues. Usually, a single ameboid cell arises from each dark spore. But, when water is plentiful and drowning imminent, the emerging amebas, known as myxamebas, simply sprout a pair of undulipodia and swim away. Such swimming myxomycote cells, former amebas, are called swarm cells from *schwarmer*, German for swimmer. The swimming mastigote stage presumably dates back to the free-living undulipodiated ancestors of these organisms. Two swimming cells, two amebas or one ameba and one swimming cell, mate. Unrestrained mating leads the fused cell to grow into a plasmodium.

Depending upon the species, sexually produced plasmodia form one of three main types of spore-bearers. In one case, a broad base which makes contact with the wood or forest floor develops. As the spore stalk grows, throbbing "plasmodial veins" shorten and thicken, and the ceaseless streaming, the characteristic internal movement of the sexually produced body, comes to an end. The body dries out, releasing spores that germinate into amebas which grow sexlessly by mitosis to form sexless offspring amebas. We may say that the plasmodium "dies." But really, it alternates between individual sexless protoctist cells and a combined, sexually produced body—making the declaration of such a simplistic cut-off point seem arbitrary. Disrupting and at the same time continuing identity, meiotic sex takes us out of ourselves, making us more—and less—than we are on our own. Once walled spores are formed and dispersed by the wind, the myxomycote only returns to its former plasmodial state when the sexual, swarm cells sprout undulipodial tails and find each other by pairs, by threes—indeed, sometimes even by dozens—and fuse again. Mating is orgiastic.

When select protoctists evolved into the ancestors of animals, plants and fungi, their reproduction, their very existence became dependent upon repeating acts of fusion. These acts of fusion probably originally occurred in the aftermath of desiccation and starvation ameliorated by cannibalism. The cyclical fusions were themselves brought on by seasonal environmental scarcity. When spring rains and summer heat resupplied nutrients, however, the waiting propagules or the fused cell masses grew and propagated wildly. These propagations, molded by natural selection, became the cyclically appearing forms we know today as bodies. The bodies were of the weird protoctists themselves and their more familiar descendants—the plants, animals and fungi. Slime molds attest to the wild sexual experimentation of bodies evolving.

SELFLESS GENES | EVER SINCE DARWIN, evolution has been pictured as a competition among individuals. Natural selection weeds out the weak individuals and leaves only the strong ones to survive. Academic neo-Darwinism, unfortunately, has been based on a too restrictive notion of the individual. Which is the individual slime mold? The throbbing mass? The frantically fusing amebas who routinely sacrifice their own individuality as they sexually merge to make a slimy whole? The cells that come together to make a whole may be genetically related, like the fusing slime mold swimmers. But often, the "individual" is composed of fused components that are genetically distinct.

The Sonoran desert termite protects great populations of the filamentous bacterial spore former, *Arthromitus chaseii*, in its swollen intestine. In fact, the single individual termite houses an entire wood-digesting microbial community. Caribbean corals eagerly ingest and retain yellow photosynthetic dinomastigote symbionts of the species *Symbiodinium microadriaticum*. Without these photosynthetic helpers the reefs themselves could not form. Neither weak dinomastigote nor strong coral animal survives as an individual on its own.

Organisms like these enhance their survival by joining into alliances. Even individual animals fuse quasi-sexually to produce new kinds of "individuals." Army ants (*Eciton burchelli*), for example, numbering 200,000 or more at a time, march through the Central and South American forests devouring leaves. They fashion their nests from the interlinked bodies of their own colony members.[6] Aggregation—of gametes and bodies—leads to larger organized entities. Aggregation into groups of both strangers and relatives enables powerful new evolutionary advantages. Evolving over time by social and interspecies interactions, "individuals" appear at ever more inclusive levels of complexity and organization.

We saw in Chapter Two that bacterial hypersex led to the first protoctists. The idea that protoctists evolved from permanent fusions of bacteria has been proven by much research including electron microscopy and molecular biology. Sexuality requires fusion. Is an egg an individual? Can a single person reproduce? The dictum that "evolution only works on the level of individuals" must be rethought. Such a principle of "the individual as the unit of selection" might make sense if only sexlessly reproducing creatures had strutted across the evolutionary stage. But all large organisms are composites. That visible beings evolved from the fusion and multiplication of microbial bodies refutes the individual selection abstraction, however convenient it might originally have been or still be in limited cases.

Furthermore, as recent research has repeatedly emphasized, animals often travel in groups to decrease their chances of dying before they have a chance to reproduce. They enhance the probability of eating well by ganging up to hunt, or to forage or to protect their young. Selfishness does not depend only upon genetic relatedness, although sociobiological oversimplification might lead one to believe so. Even the simplest protoctist "individual" evolved hypersexually—from a permanent symbiosis of quite unrelated bacteria. Coming together, power in numbers, has been crucial in at least two major evolutionary transitions: the hypersex that led from bacteria to nucleated cells, and the fusion sex that led to large bodies such as those of slime molds, sea weeds and, later, plants and animals.

The concept that organisms transcend themselves by living together—with conspecifics or genetically unrelated strangers—has, in mainstream zoology, been disparaged for decades. Attempting to be scientific, population biologists and especially zoologists have treated organisms as if they were physically isolated particles. In neo-Darwinian formulations, organisms, or even just their genes, are indivisible, independent and self-interested units. But this notion of genes and animals as individual atoms is destructive. Organisms are open, growing systems with many opportunities to link themselves temporarily or permanently to other such systems.

WHY SEX? | INDIVIDUAL LIVE ORGANISMS, in fact, are not atoms—or any other kind of particles. They are not even things. Living beings are bounded, thermodynamically and informationally open processes. Their boundaries are always changing. Through their membranes, skins and orifices they connect with their surroundings and with each other. Transforming energy and producing entropy, each individual organism both maintains and, if it is sexual in any way, merges. Whether the transgenic sex of bacteria, the hypersex of protoctists or the fusion sex of animal, plant or fungus, all sexual beings both maintain themselves and merge. Organisms are far less independent individuals than modern neo-Darwinian biology has assumed. Indeed, merging, at the bacterial and later the eukaryotic sexual level, has been directly responsible for major evolutionary transitions. Dying protoctists saved themselves by sexual fusion. Later, as they evolved to adjust to cyclically improving environmental conditions, they went on to grow in their fused state to form the first bodies. Sexual fusion, imperative in animals, takes us out of ourselves and spurs new forms of social organizaion.

Once cycles of fertilization following meiosis became established, they flourished. Why? Details involving the retention of sex depend upon the details of the organism lineages in which that sexuality appeared in the first place. Because most sexual examples are drawn from animals, whereas most sexual diversity is in nonanimals, the literature tends to be confused. The history, meaning and level of complexity of sex differs to such a degree in different groups of organisms that any general theory of sex is bound to be fallacious.

One especially recurrent but dubious reason given for why organisms retain sexual fusion touts sex as some sort of genetic rejuvenating mechanism. This assertion is based on observations in certain ciliates (*Paramecium aurelia*). Growing by direct division these paramecia reproduce mitotically in the absence of any sexual liaisons. Such paramecia populations survive for only months, while their relatives that indulge in sexual conjugations survive indefinitely. But is it the sex that rejuvenates? Not necessarily. A paramecium that prepares for sexual conjugation but finds no partner experiences the "self-mating" process known as autogamy. The diploid nucleus of this single cell divides by meiosis to produce four haploid offspring nuclei. These nuclei from the very same cell fuse with each other in the complete absence of any sexual partner. Mitotic reproduction follows. The ensuing, entirely inbred lineage of paramecium survives just as long as its conjugating relatives that underwent two-parent sex. And such self-fertilization renders the ciliate entirely homozygous: genetic variety— produced in sex and often cited as the reason organisms retain sex when they can reproduce faster without it—has been reduced, not augmented. Yet the paramecium is recharged and rejuvenated and is again capable of sexless reproduction for generations.

Clearly, something more than mating or the receipt of genes via sex is crucial, at least to paramecium rejuvenation. Not new genes but meiosis, the complex process of reduction of chromosome number per cell saves the lineage from dying out. We suspect, then, that meiosis, followed by fusion, even inside the same single parent, is the imperative. Meiosis and fusion at the cell level, and not necessarily two-parent sex, is the key to rejuvenation.

Meiosis, followed by fusion of nuclei from either one or two parents, is repeated whenever a new animal or plant body develops. Meiosis and fertilization are required by sexual organisms for their being, maintenance and growth. Sexual protoctists, plants and animals are complex, incessantly metabolizing bodies in whose history sexual cell fusion and meiotic relief of diploidy are deeply embedded. Thermodynamically and genetically, meiosis reestablishes the initial conditions from which bodies, those intricate dissipative structures, grow. Meiotically sexual organisms—mating protoctists, most zygo-, asco- and basidiomycotous fungi and all plants and animals—must "go back to the drawing board." Nuclear fusion begins the developmental process, and meiosis relieves it so it can begin again in the next generation. The more complex the body, the greater the number and diversity of its integrated parts, the more stringent, apparently, is the requirement to "start at the beginning" by fusing and relieving fusion by meiotic reduction division.

Self-repair and regeneration capabilities diminish over evolutionary time as bodies become more differentiated, and complex. Myxomycotes endure the mincing or removal of large sections of their plasmodia, yet they continue, unfazed, moving and feeding. They increase in size by the strange expedient of joining and fusing either entirely to other single cells or to entire plasmodia. Aquatic animals such as polyps, starfish and flatworms regenerate most of their bodies from severed sections. Crabs, lobsters and many insects regenerate lost limbs. Even so-called "higher" animals[7] such as crippled lizards and salamanders can regrow lost legs. Primates and carnivores such as ourselves, our dogs and our cats, however, renew only select tissues and heal only limited injuries. We can grow back our hair but not our head. The crux of our existential crisis is that to totally regenerate ourselves, we must have sex and die as conscious individuals.

The need for brown algae, many water molds and diatoms, most fungi and plants and animals to return to their ancestral fusions each generation is reminiscent of a child's need to start at "A" to recite the alphabet. The cyclical nature of fusion making doubleness and meiosis relieving this state can be seen as a sort of condensed space-time that unfolds in the development of bodies. Historical systems, far from equilibrium, differ from near-equilibrium systems in that they do not tend toward inevitable dissolution. Rather, they maintain their "selves" by cycling series of actions that derive from discrete histories. Pathways were established and, since they were incorporated into the living being in question, they can be revisited. Both physiology and memory retain in living matter the peculiar details of a given historical trajectory. The history of living systems unconsciously reappears. History is stored and perhaps forgotten by living matter but it does not disappear. Ritualized sexual fusion followed by meiosis originally evolved because it protected the ancestors from untimely death. Today dependent paths of history must be retraversed. Compared to nonliving systems, organisms are so extremely ornate because they embody eccentric, multibillion-year-long histories.

In meiotic sex the recapture of initial conditions that leads to the growth of complex bodies always requires that essence of fertilization, nuclear fusion. That fertilization occurs at some point in the life cycle is ensured by chemically mediated sensations in beings of complementary gender. Eventually meiosis and cyclical cell fusion became nonnegotiable for survival. Meiosis occurs even in the many protoctists, plants and animals which bypass two-parent sex. Some lineages that practice multigender or two-parent sex have lost two parent sex in certain strains or species. But meiosis and nucleus fusion persist in the single parent. This observation tells us that meiosis and fertilization are as important for the development of individual beings now as they were for the origin of these beings in the past. Within the borders of naturally occurring variation, each species guarantees, by repeating its sexual life cycle, that its own highly specific form of gene-directed material organization will be passed on. Sex "locks in" the precise form of a body, the living organization we recognize as a given species of sexual protoctist, fungus, plant or animal. Our kind of protoctist-based "vertical" sex, where fused nuclei divide and pass to the next generation (as opposed to the rampant gene-transferring "horizontal" sex of bacteria in Chapter Two) is essential to the self. The beginning of the cyclical order preserved by meiosis is mating and cell fusion. The end of the cyclical order preserved by meiosis is aging and death.

Meiotic sex and tissue-level multicellularity, the hallmarks of the two great groups of embryo makers (plants and animals), both evolved in protoctists. Certain sexual water molds, called chytrids, develop when two swimming spermlike mates fuse. These organisms were probably ancestral to the never-swimming fungi. The sexual tissue-forming green algae (chlorophytes) were probably ancestral to the plants. The little sexual protoctists, such as the zoomastigotes, were probably ancestral to *Trichoplax*, sponges and other early egg-sperm, blastula-making aquatic animals. Animals appear at the end of the Proterozoic eon well before 540 million years ago. The evolution of meiosis, which in animals, plants and fungi is correlated with complex cell and tissue differentiation, must have preceded this date. Animals, plants and fungi return every generation to a single fertile diploid nuclei. In us, we return to the fertilized egg. Meiosis, especially, the chromosomal DNA-alignment process in the stage called prophase I, may be a sort of "roll call" ensuring that sets of genes, including mitochondrial and chloroplast genes, are in order before development of the plant or animal embryo ensues. We saw that one-parent meiosis sufficed to rejuvenate paramecia. Meiosis is probably maintained in all fusion-sex organisms because the process itself is essential to the physiology of development and survival. When complex meiotic organisms reproduced, meiosis, the prerequisite to fertilization, had to be retraversed.

The ideas of sexual origins we present here differ from commonly taught assumptions on the role of sex in evolution. Bacterial sex is often ignored while two-parent sex is considered to be a crucial evolutionary advance. The evidence, however, suggests that bacterial sex is crucial while two-parent sex is never directly selected. Bacterial sex, a modified DNA repair mechanism, allows organisms to accept new genes as easily as we catch cold. The bacterial legacies of enzyme-mediated DNA repair reappear in meiosis prophase I of animals and plants. Without bacterial sex processes the meiotic sex of animal and plants never would have evolved. Thus, although we see the DNA repair processes of meiosis as crucial, the two-parent two-gender aspect of meiotic sex came later. In animals, it is an evolutionary legacy. Entire animal taxa, for example the rotifer phylum, with its hundreds estuarine, marine and fresh water species, bypass biparentality. These animals show no loss at all of evolutionary success or of variation. What was selected for were the complex tissues and organs, not the sexuality, of these animals. These rotifers and other animals that do, in so many cases, forego two-parent sex, did not abandon meiosis. They could not have.

Might human males, like rotifer males, become evolutionarily redundant? Certainly, the cloning of women's eggs could, in principle, circumvent our two-parent sexual cycle. But it is doubtful that meiosis and fertilization in such women can be entirely abandoned. Rather these hardy mothers will be self-fertilizing. Their haploid eggs will probably require a fertile boost of self fusion—the egg nuclei fertilized by an egg equivalent such as another haploid nucleus from the female's own body. Indeed, this is what occurs today in all-female rotifers.

OFF WITH THE RED QUEEN'S HEAD | EVOLUTIONARY BIOLOGISTS who ponder the question of why so many organisms are sexual like to point to rare species—really only populations—of whiptail lizards and all-female rotifers which avoid the bothersome task of sexual reproduction prevalent in the "normal," heterosexual, species.[8] Why, they ask, have not female-only "species" replaced their courting neighbors? Twice as much reproduction can ensue, on average, if males (who, in principle cannot reproduce alone) are replaced by females all of whom can, in theory, reproduce. The textbook answer is that sexual organisms, intrinsically changing, are better adapted to rapidly changing environments. But both the textbook question and its ready answer mislead.

Despite their apparent "loss of sex," their lack of males with which to fornicate and their female-on-female "pseudocopulations," all-female populations of whiptail lizards in the southwestern United States retain an underlying sexuality. Their cells undergo meiosis, fertilization and other normal two-gendered processes. Sex is far from completely lost. These lizards are not asexual; instead, they are single-gendered and uniparental. Here we see the problem of considering sex as if it were a single process: the loss of one gender, the male, is entirely different from the loss of sexual cycles of meiosis and fertilization! Inside the ovaries of these whiptails, female haploid cells fuse to form female diploid cells capable of development into lizard embryos.

Let us look more closely at the standard argument. Because time and energy must be expended in searching for a mate, biologists have assumed that organisms would be selected to reproduce without sex because this would avoid wasting time and energy. Natural selection must lead, by this logic, to beings who reproduce themselves simply by standard mitosis: cell divisions, in the absence of sex, that lead to the budding off or equal division to form new bodies. In the economic language zoologists use (or—from our viewpoint—misuse) to describe evolution, the "cost" of sex—finding mates, producing special sex cells with half the usual number of chromosomes, and investing time in these activities—seems all out of proportion to any benefit. To resolve this apparent dilemma certain biologists reason that sex itself must confer an advantage. One favorite suggestion is that sex remains because it increases variation in the offspring. This variation, it is claimed, allows sexual organisms to adapt faster to changing environments than do their sexless counterparts. Because organisms themselves—parasites, predators and competitors—are crucial parts of the environment, it was further reasoned that sexual diversity would beget increased sexual diversity. Such thinking laces many textbooks and popular articles and has even has acquired a catchy name: The Red Queen Hypothesis. The Red Queen in *Alice and Wonderland* ran,

> ...so fast that it was all [Alice] could do to keep up with her: and still the Queen kept crying "Faster! Faster!" but Alice felt she *could not* go faster, though she had no breath left to say so. The most curious part of the thing was, that the trees and the other things round them never changed their places at all: however fast they went, they never seemed to pass anything...." In *our* country," said Alice, still panting a little, "You'd generally get to somewhere else—if you ran very fast for a long time as we've been doing."
>
> "A slow sort of country!" said the Queen. "Now, *here*, you see, it takes all the running *you* can do, to keep in the same place. If you want to get somewhere else, you must run at least twice as fast as that!"[9]

The Red Queen Hypothesis that sexuality produces greater biodiversity is probably not even a hypothesis since it is based on the idea of sexuality as a single process. The insistence on the importance of sex to produce variety to keep up with rapidly changing environments is in inverse proportion to the data. Sexual rotifers are not more varied and adaptable than uniparental ones of all-female populations. The evidence is against the notion that sex-fomented diversity is required to survive and evolve in changing environments. When Harvard biologist Matthew Meselson used molecular biology to compare related uniparental (asexual) and biparental (bisexual) rotifer species, he found that the all-female rotifers showed 300 times as many genetic differences in their alleles![10] Hundreds of species of bdelloid rotifers have been described in which no one has ever witnessed a sperm cell, a male or even an act of fertilization. The very abundance of single-parent (all-female) rotifer populations attests to a great deal of heritable variation. One-parent organisms seem to be able to generate plenty of measurable variation for natural selection. Additionally, spontaneous generation of rapidly varying surface antigens occurs on the membranes of both *Borrelia* (the Lyme disease spirochete bacterium found in ticks) and *Trypanosoma* (a tropical protoctist, a blood-disease parasite). Generation of rampant drug-eluding surface variants of these organisms shows that, in the absence of any sex at all, reproduction, far from always ensuring identity, can generate plenty of heritable variation. Without any sex—transgenic genetic recombination, sexual fusion or meiosis—variation abounds. The Red Queen idea is simply a cute name for a zoological myth.

Many processes, in fact, such as mutation, symbiogenesis and even physiological stress, are capable of generating variation, measurable differences in organisms. Sexual breeding is also one of these generators of variation. But the crucial experiments of quantitative comparison between the same type of populations of organisms growing with and without two-parental sex under identical environmental conditions have not been made. Indeed they *cannot* be made since the vast majority of sexual species cannot be induced to lose their sex and live. Thus any statement that sex augments variation relative to lack of sex is unfounded. When, in cases like the protoctists and the rotifers, biparental sexuality can be experimentally manipulated, environmental conditions must be altered to suppress the sexuality. In conclusion, then, the sexual production of variation is not directly comparable to variation generated in other ways. The fertilization-meiotic sexual imperative remains an inescapable part of the life of large organisms. The cyclical, gradient-breaking path of protoctists that evolved into the first animals and plants was inescapably bound to their sexuality.

THE MEIOTIC IMPERATIVE | OUR EXPLANATION of why sexual individuals trouble themselves to seek out mates of complementary gender only to dilute their genes is simple. They have no choice. To survive the winter cold or summer drought they must undergo sex. If they are to participate in the solar gradient-breaking process, they must indulge in sexual fusion.

All the ways in which sex intertwines in the existence of animals are not known. Sex is not, as too many biologists assume, directly selected. It is, rather, part of the path by which we complex eukaryotes came to be what we are. Biparental mating, meiosis and fertilization are deeply embedded, dating back to a time when cells, by doubling (fertilizing) and breaking up (dividing) on a seasonal basis, evolved cycles of multicellular growth and reproduction. Usually indispensable to forming the new fertile animal egg that becomes the blastula, animal sexuality is never entirely lost. Two-parent sexual activity is so tied to meiotic processes that it is never easily discarded in animals, especially mammals. Plant, fungal and protoctist genders are far more labile. Many profound variations on fusion sex have survived. The fundamental reason for these differences is that other avenues of reproduction are open to plants, fungi and all protoctists (e.g., duckweed that grows by mitosis without sexual flowering, *Penicillium* mold that disperses by mitotic production of spores, and so on).

Although multicellular and sexual, the earliest chordates—our same-phylum animal ancestors—had not yet evolved to deposit calcium phosphate inside their bodies as a structural support in the form of a skeletal system. Yet, they did have a notochord, the cartilaginous support extending down the back from head to tail that would later be replaced by the spinal column and its nerve fibers connecting the genitals to the brain. Ancient chordates no doubt ejaculated sperm and produced yolk-filled sedentary eggs.

Attracted by specific chemical cues, the male of any given species perceives some females of its own (but not other) species as sexy. Often bombarded with a surfeit of sperm from many males, most eggs let only one sperm cell enter and fuse with its nucleus. To be an organized, tissue-differentiated animal requires such sexual discrimination every generation. Animal multicellularity is built on two-parent sex. In your ontogeny—your development from a fertilized egg, a zygote, to a mature adult—you and all your siblings travel the ancestral road and stop at the ancestral inn of sex because that is where all your ancestors had to stop in order to survive. With no inn on the well-trodden road, the dangerous trek is not possible.

Animalhood, in other words, is path-dependent: animals depend upon repeating particular events in the history of their ancestors. Today, no armadillo, gecko or infant is born or will survive without moving through the processes of meiosis, gender formation and fertilization. The cyclical road to modern human beings requires traversal of the loop-to-loop of sex before returning to the "starting point" of coupled sperm and egg. Except our red blood cells, which lack nuclei and therefore have no chromosomes, and our sperm or ova, which have only one set, nearly every cell in our bodies has at least two sets of chromosomes. Each of our cells combines our mother's and our father's nuclear inheritance. We are meiotic and sexual to the very core of our being.

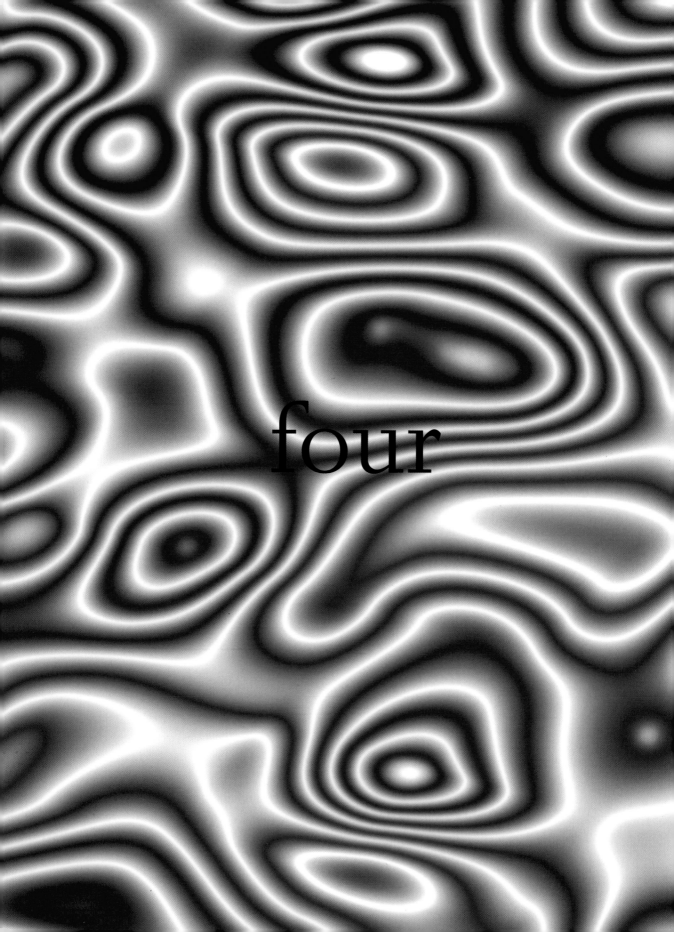

four

THE KISS OF DEATH: SEXUALITY AND MORTALITY

For sale anarchy for the masses;
irrepressible satisfaction for rare connoisseurs;
agonizing death for the faithful and for lovers!

— Rimbaud

THE SEX/DEATH CONNECTION | THE EVOLUTION OF SEX was like a pact with the devil. Fusion sex, fertilization followed by meiosis, allowed beings to survive the cycles of the seasons. Sex let animals grow elaborately complex, multicellular bodies from fertile eggs. But the price for ecstasy beyond identity—the sexuality that impels us to unite with each other and make a new being beyond ourselves—was high. Sex, at the cellular level, has been connected to death for perhaps seven-hundred million years. The parent bodies had to die. The evolution of sex in animals was accompanied by the aging unto death of their bodies. Death on cue, death on schedule, so-called programmed death was part and parcel of fusion sex since its single-cell beginnings. The romantic connection of sex to death in art reflects real evolutionary history.

Strange as it may seem, the aging and death which we consider normal—and which give us so much grief when we consider our ultimate demise—did not exist at the origin of life, or for millions of years thereafter. Aging, decline and death, as we know them in mammals, first evolved in ancient protoctists like the ancestors to ciliates—sexual one-celled microbes. The minute single cells had to sexually fuse at least their nuclei to survive. Sexual fusion reset the clock of life and forestalled aging. Descendants of these microbes evolved animal bodies capable of speciation. [**PLATE 31**] Our entire conscious life, from childhood on—including our brain-based consciousness of our own impending death—is a latter-day result of the growth properties of protoctists that came together in sexual embrace. The clonal growth of fused nucleated cells that created cysts and other resistant progagules unleashed many new opportunities for life. All animals, and most plants and fungi, represent growth patterns established by protoctists that grow, however briefly, in a fused (diploid) state before they return to their ancestral unfused (haploid) state. All plants, notably mosses and ferns in which the haploid and diploid stages are separate little plants, grow their bodies—called gametophytes—in the unfused state. This would

PLATE 31
Karyotypic fissioning: a way in which mammals form new species. The protoctist legacy of centromeric premature replication leads to speciation in hoofed mammals, lemurs and others. (See note 6 of chapter 4.)
[*Kathryn Delisle/José Conde*]

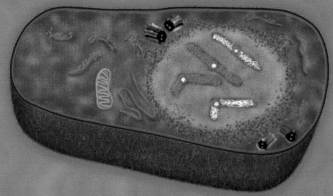

unfissioned parent

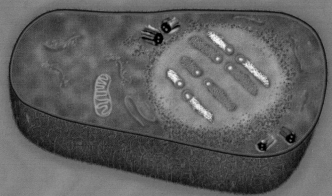

fissioned offspring

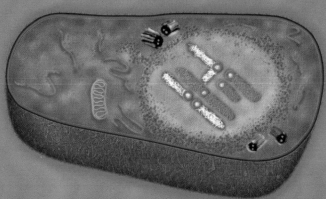

hybrid

be comparable to the sperm of a man and the egg of a woman growing all by themselves into little one-chromosome-set eaters and excreters—a sperm or an egg growing all by itself into a multicellular individual. Growth in the unfused single (haploid) state always alternates with the fused doubled (diploid) state in sexual organisms. Nonnegotiable return to the ancestral condition implies that the fused state can only be temporary. For humans, what grows in the fused state is nothing less than our body with its brain capable of understanding our future demise and challenging us to search for meaning in our lives.

A long cultural history associates sex with death. Influenced by Hindu ideas on the fundamentally deceptive nature of reality, including biological reality, German philosopher Arthur Schopenhauer (1788-1860) held that the attractiveness of the opposite gender was a fundamental illusion designed to perpetuate what he called "the genius of the species." As such, all gossip about who was, should, would or could be mating with whom was in fact a meditation on the genetic composition of the next generation. We care as much or more about sex, romance and our children than about ourselves probably because our brainy bodies are destined to die, whereas our genes, in their protoctist-like sperm and egg containers, possess immortality. As China-born physicist, John Dobson, founder of the Side-walk Astronomers and a student of the Upanishads (the philosophical sections of the Vedas, or ancient Sanskrit scriptures), puts it:

> The prime directives of the genetic programming are to direct a stream of negative entropy upon ourselves and to pass on the genetic line. That is why we feel ourselves to be the doers of action and the enjoyers of its fruits. It is just a genetic mirage. The genes have us persuaded that by following their dictates we'll reach the peace of the changeless, the freedom of the infinite, and the bliss of the undivided. They don't have it to give. We don't get the undivided; we get a family. You must have noticed.[1]

The German poet-scientist Johann Wolfgang Goethe's (1749-1832) "The Sorrows of Young Werther," a story about a man who commits suicide in the wake of his hopeless infatuation with a woman he cannot possess, helped to touch off the romantic movement and inspired young people in the German countryside to kill themselves in emulation of their romantic hero. The Austrian physician and founder of psychoanalysis, Sigmund Freud (1856-1939), stressed the unconscious importance of sexual imagery to early childhood. At one point in his career Freud traced all psychological activity to two great drives, eros, the sex drive, and thanatos, the death drive. The sex/death connection is also featured in the biblical story of Eve who yields to temptation from the devil in the form of a serpent, thus bringing about the first couple's eviction from the Garden of Eden and the fall of humanity, through original sin, from graceful immortality to earthly lust and sex. From the ancient story of Genesis to the latest spy movie's femme fatale, sex and death are intimate.

This is no coincidence. The association of sex and death in the popular imagination reflects life itself. Sex and death are tightly interwoven as an evolutionary legacy of the ancient merge-for-survival strategy of haploid microbes. The conservative nature of evolution has linked sex and death in the descendants of such microbes from the earliest times—before plants, fungi and animals evolved from them. Sexually reproducing organisms exist as discrete individuals for only a limited time. The sex-death connection does not exist in bacteria or even amebas, euglenas and other nonsexual protoctists. Reproducing by fragmentation such protoctists are, in principle, immortal. None of the estimated 250,000 species of extant protoctists require sexual fusion to reproduce. Many, such as the brown and red algae, rely on sex to invoke developmental changes that allow tolerance to seasonal extremes. Most protoctists engage in sex only to survive. When the environment, through natural seasonal alterations of sun and shade, heat and cold, wet and dry, threatens, protoctists respond: they develop complementary genders, sometimes not two but dozens of them, which attract each other.

Single-parent reproduction without sex—both bacterial fission and, later, protoctist mitosis—was the norm for the first two billion years of life on Earth. The rampant, creepy growth that we find freakish when it appears as cancerous tumor, pus-filled wound or bacterial infection is in fact life's original sexless modus operandi. Such rampant reproduction and growth by cell division was interrupted and belatedly improved upon with the innovation of sexuality. Parts of cells were made redundant and unnecessary by the doubleness of fusion. Death became seasonally inevitable. Detailed self-destruct or preferential growth pathways evolved that regulated the number of chromosomes, chloroplasts and mitochondria per cell, ensuring double-monster fused cells' survival. Bacterial cell division and mitotic cell division—the unfused state—were and are the staples of growth and reproduction. The infrequent cyclic fusion which began haphazardly had to be controlled because of its potentially monstrous consequences. Ultimately it became the stark facts of aging and programmed death.

Provided with sufficient energy, food, water and space, all bacteria and many protoctists are still immortal. They metabolize and grow without limit. They double their DNA and other cell components and divide into two offspring cells. In its essential thermodynamic simplicity, life expands by an unfettered cell division urge. There is no intrinsic corpse, no mortality, no necessary "being towards death," as the German philosopher Martin Heidegger (1889-1977) put it. Two cell beings appear from one. If the offspring cells remain together, they may form the beginning of a body or tumor. If, after division, they go their separate ways, we call it cell reproduction. Here, again, no death is detected: no dead body, no cadaver forms. Instead, in a rather seamless process, two new beings form where before there was only one. Bacterial binary fission or protoctist mitosis without cessation is the underlying fact of evolutionary science: the impetus to grow and reproduce is our absolute microbial legacy that resists restraint. When the single parent cell generates two cells, the former parent is not dead but divided. By contrast, sexual "individuals"—the protoctist legacy of bodies that grow, mature and inevitably die—evolved relatively recently, less than a billion years ago. The death of an entire individual organism, programmed and predictable, evolved not in bacteria, but in their successors, the protoctists. All protoctists reproduce—at least sometimes—by direct one-parent division: mitotic cell reproduction. Some, however, but by no means all, survive seasonal hostility through the fertilizing fusions of meiotic sex. We count these among our ancestors.

ANCIENT CELL DEATH | THERE ARE TWO kinds of cell death. The first, evolutionarily older, is called cytocide or necrosis. This is avoidable death. Cytocide is accidental, externally caused. [PLATE 32] The second, apoptosis, is death from the inside, death on the installment plan, inevitable death. Apoptosis, the cellular equivalent of suicide, probably proliferated due to selection pressures on sleek or luxuriant multicellular bodies shaped by growth and death. When an embryo enlarges and develops distinctive parts, this is driven not only by growth and mitotic reproduction of its cells but also by programmed cell death. The grim reaper is also a happy shape. [PLATE 33]

The first cells were not scheduled, genetically, to die. In cytocide a cell struggles mightily to continue to function. Resisting death, it attempts to maintain itself against the forces that would kill it. Human body cells threatened by external insults violently resist the tragedy of cytocide. A human cell irreparably damaged by toxins or lacking nutriment from the surrounding lymph or blood will soon die by cytocide. Starvation, desiccation and other deprivations may lead to lack of energy needed to pump potassium ions in and sodium and calcium ions out. Water rushes in through the cell's membrane and the deprived cell explodes in the violent

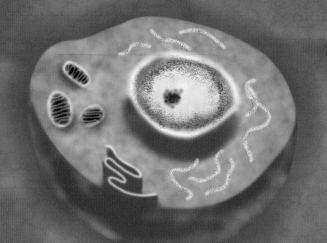

Cell disintegration
begins.

Nuclear membrane
pops open.

Cell bursts open as
membranes dissolves.

imposed death of cytocide. Such death is unnecessary. If provided enough food or enough of the energy compound ATP on time and at the correct location, the cell would have survived. Maintenance of cell membrane integrity requires incessant vigilance of membrane repair and ion pumping. A cell that dies by cytocide attracts the macrophages—Greek for "big eaters"—of the immune system. These are police-like cells that monitor, engulf and drag away the remains of the victims of cytocide. Macrophages induce nearby cells, called fibroblasts, to seal up lesions (sometimes visible as scar tissue). Cytocide is an unpredictable cellular emergency—inducing responses from neighboring survivors.

Apoptosis is different. Apoptosis is programmed cell death. Unlike the murder of cytocide, apoptosis is suicide of a strange sort. It is suicide without option—inevitable, natural and necessary suicide upon which sexual bodies depend for their normalcy. Cells, as components of all sexual organisms, enjoy natural limits to their growth. Most divide only for a discrete number of generations and then stop. These cells that actively destroy themselves in a predictable fashion at certain moments during a life's history are called apoptotic. Like the "negative space"—the empty area around a model on which an art teacher has her students focus—the cells disappearing by cytocide shape the organism that remains. In the human embryo, for example, the millions of cells that form webbing between the toes of the developing fetus undergo programmed death.

PROGRAMMED CELL DEATH | IN APOPTOSIS, cells are never killed by some force from the outside. The nucleus degrades and the DNA it contains gracefully falls apart. The term itself comes from a Greek word meaning "to drop away," as in the graceful falling of petals from a flower. Microscopic preparations of apoptotic cells reveal that the demise occurs with stately precision. Apoptosis is serene. Instructions for death are sent out by the DNA of the nucleus into the cytoplasm. The RNA message is converted

PLATE 32
Cytocide: violent and resisted death in mammalian tissue cells.
[Kathryn Delisle/José Conde]

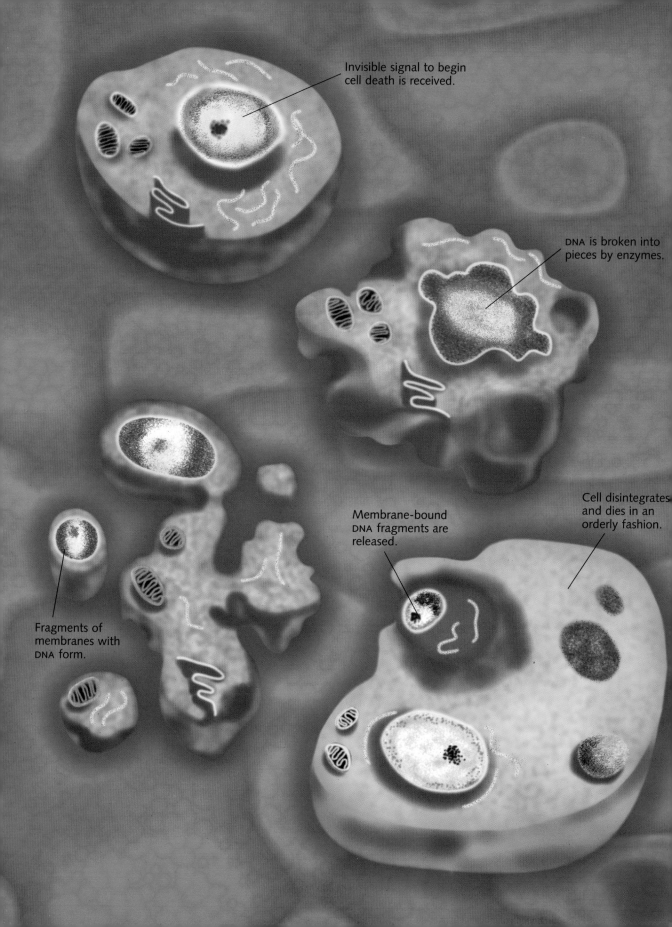

Invisible signal to begin cell death is received.

DNA is broken into pieces by enzymes.

Membrane-bound DNA fragments are released.

Cell disintegrates and dies in an orderly fashion.

Fragments of membranes with DNA form.

into proteins, enzymes which, in turn, begin to cut the DNA into small fragments. Genetically shredded, the DNA can no longer be read. Once cleavage of the long strand of DNA into small pieces begins, there is no return: the cell will die. Like a coma patient who lacks electrical activity in the brain, the cell may continue to live for awhile, synthesizing proteins, but all hope is lost.

Apoptosis evolved in the same organisms that invented meiotic sex, the protoctists. Today the old macronucleus of paramecium gracefully dies during autogamy and sexual conjugation, making way for a new one to develop. Trypanosomes—tiny protoctists requiring food and energy from the blood or other tissue of the large animals in whom they live—exhibit fully developed apoptosis. One of the first signs of ensuing programmed death in these minuscule swimmers is a rounding shape. The tiny trypanosome cell develops holes. Inside the trypanosome nucleus, chromatin (the chromosomal material) begins to clump. Methodically cut pieces of chromosomal DNA, chopped by specific enzymes, are released into the surrounding fluid. To survive, normal swimming trypanosomes, often carried by insects, must attach to the tissues of the animals they infect. Then those trypanosomes ejected in the insect's bite proceed to grow by division inside the victim. Many trypanosomes, however, fail to contact and infect their animals. Unconnected trypanosomes commit cell suicide by apoptosis. That such trypanosomes enjoy complex differentiated life stages inside insects and mammals and display fully developed apoptosis, yet lack meiotic or any other fusion sex, suggests to us that apoptosis evolved before fertilization-meiosis mating cycles. Apoptosis, we suspect, evolved in early protoctists responding to cyclical environmental changes even before meiotic sex took hold.

Although present in other species, the process of apoptosis is best known from studies of mammals including human tissue. The apoptotic cell in animal bodies first detaches itself from neighboring cells. The

PLATE 33
Apoptosis: programmed cell death. The dying cell (upper left) peacefully breaks up its DNA into membrane-enclosed vesicles (lower left), which disperse to be engulfed by other cells (lower right).
[*Kathryn Delisle/José Conde*]

bubbling membrane undulates and then breaks apart and falls away into fragments. Severed cell bits, called apoptotic bodies, contain ribosomes and other organelles such as mitochondria. Surrounding mammalian cells in tissues do not respond to local apoptosis as they do to the emergency of cytocide. No alarm calls. Tissue spaces do not enlarge to surround the apoptotic cell as they do in the cell frantically dying by cytocide. No tissues fill with white blood cells that release messenger chemicals to the immune system. The tissue in which the cells undergo programmed death remains calm. Inflammation does not occur. The membrane-bounded apoptotic bodies are smoothly recycled into living matter; the DNA and protein parts broken down are used for resynthesis. The neighboring cells, including resident macrophages, engulf and digest the nutritious remains. The tissues in which apoptotic cells quietly die remain healthy. Programmed death, like shedded hair or menstruum, is natural. Bodies—whether animal, plant or fungal—require programmed death to live.

SELF-DESTRUCTION | SMALL, SQUIGGLY AND TRANSLUCENT nematodes with the species name *Caenorhabditis elegans* are common garden roundworms. Fortunately for scientists, they have a cuticle, a covering through which muscle and nerve cells can be seen with a microscope while the animal is alive. Not surprisingly, they are one of the most well-studied animals on Earth. The adult nematode, a millimeter long, ingests bacteria as it wriggles. Composed of exactly 959 cells, the adult worm grows from a fertile egg in three days. As the nematode matures, an additional 131 nematode cells appear, only to die during the transition to adulthood. They die by programmed cell death.

Despite the evolutionary imperative for survival, many animals, when they become too crowded, naturally decimate their own populations. Such self-destruction at the individual level seems to enhance survival at the social level. Density-dependent behavior has been described in mammals.

Crowded rodents, for example, are more likely to engage in "gang" violence, homosexuality and self-mutilation. The female mouse (Mus musculus) fails to implant fertilized eggs in her uterine wall if she is exposed to the odor of a strange male—an indication of increased crowding. Meadow voles spontaneously abort under similar conditions. Resisting growth has proved an excellent means of ensuring future growth potential. Apoptosis as cell suicide, and animal behaviors that reduce population size or prevent conception, are of an evolved, life-wide strategy of moderation, delaying present gratification for future prospects.

Reminiscent of "downsizing" corporations that lay off workers, bodies naturally "let go" of their own cells or organelles by apoptosis. Calm, predictable cell suicide follows a spate of rapid cell growth. Tissue wounds repair, organs develop as supernumerary cells are removed in an orderly fashion. Past deaths inform present organization. As Austrian physicist Erich Jantsch says in his book The Self-Organizing Universe, "[O]ne important result of evolution may be seen in the increasing intensification of...life in the present by means of including the experience of the past and anticipations of the future. Biological evolution makes the experience of an entire phylum, starting from the formation of the first biomolecules, effective in the present."[2] The death of groups of cells is an absolute requirement for the maturation of a body from an embryo.

THE ETERNAL WOMB OF HENRIETTA LACKS | ANIMALS EVOLVED from immortal protoctist ancestors that evolved mortality and settled down. They still develop from immortal egg cells. Normally restrained, stressed body cells sometimes revert to their ancestral growth mode by becoming cancerous. Cancer cells are throwbacks to the earlier lifestyle of unrestrained mitosis. The carefully differentiated community which is the animal body tends to disintegrate. Individuality is maintained only by incessant reiteration and reassurance, as most of us realize intuitively.

We animals exert tenuous, unconscious control over our tissues and organs, protecting them by a cooperative, instinctive immune system. The cooperation of the vertebrate immune system evolved relatively recently. Embryos display growth by mitotic cell division; they develop by intracellular and locomotory movements. Predictable cell activities—metabolism, cell growth, cell division, cell movement, cell fusion and cell apoptosis—always accompany embryonic development. Cell division slows and stops at certain locations; cell deaths ensue on cue; the embryo develops.

The body resembles an ecosystem in which the rapidly growing stages of early development correspond to territory colonization by pioneer species. Then limits come into play. Rapidly growing pioneers are soon accompanied by a wide range of other species as ecosystems enter a period of slower growth. Species diversity appears in more mature ecosystems after the growth of fast-growing pioneer species slows down. Similarly in the mammalian body, the rampant early cell growth of the embryo tapers off. Different tissues and organs appear as, in *utero*, the characteristic human form develops. [**PLATE 34**] Indeed, in animal development and ecosystem secession we recognize similar underlying processes at very different scales.

HeLa cells afford us a most vivid example of the human body's latent tendency for rampant cell growth. HeLa cells are named after Henrietta Lacks, mother of four who, in 1951, was admitted to John Hopkins Hospital with cancer of the cervix. After initially encouraging radiation treatment to the tumor, part of which was removed for study by a pathologist, Henrietta Lacks' aggressive cancer spread to nearby organs. She died in October of the same year. Since her death, however, her cervical cells, which continue to thrive, have been exported to laboratories all over the world. George Gey, a researcher of polio at John Hopkins Medical School, had been frustrated in his failure to grow polio viruses in the laboratory. As do all viruses, human polio requires live cells for reproduction. Gey was surprised and delighted when he discovered the vigorous growth and,

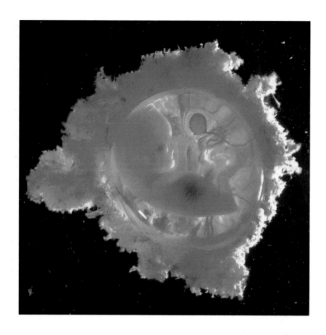

therefore, the research possibilities of the cervical cancer cells taken from Henrietta Lacks. Indeed, so aggressive are HeLa cells that they outgrow other human tissue growing in culture. A scientific scandal erupted when, in 1966, Stan Gartler, a University of Washington geneticist, found that independently isolated heart, kidney and liver cell tissues had been contaminated and replaced by the more rapidly growing HeLa cells. Medically, HeLa cells are important because they continue to provide researchers with human cells in which viruses can be studied without the use of whole human bodies.

Prior to the widespread use of HeLa cells, human cells had not successfully been grown in culture. Investigators were stymied by the so-called Hayflick number. Cell biologist Leonard Hayflick, working in the 70s and 80s, found that fibroblasts, worker cells that lay down scar tissue, divide only a limited number of times after removal from the human body for culture. Provided with nutrients, kept humid at body temperature and

---

PLATE 34
Unborn human infant.
[*Lennart Nilsson*]

supplied with nitrogen and carbon dioxide, fibroblasts can be cultured or grown in Petri dishes, where they can continue to divide. Hayflick found that whereas fibroblasts taken from a middle-aged person might divide 20 or 30 times, fibroblasts taken from fetuses divided an average of about 50 times before they stop and die. If fetal fibroblasts were, after 20 divisions, stored in liquid nitrogen for over a decade, they would, when warmed and resupplied with nutrients, begin where they left off. They divided another 30 times before degenerating and dying—showing that fetal cells keep track of their age. Each type of cell seems programmed for a specific number of cell divisions.

Cells removed from embryos are called embryonic stem (ES) cells. These cells are removed from mice at an early stage. In mice, this is when the mouse embryo is still in the oviduct, before implantation in the uterus wall, and has grown to only about 100 cells. These mouse ES cells, manipulated to prevent differentiation and termination of growth, grow indefinitely in culture. Cells growing in culture, when added to a second blastocyst, are accepted into the mouse embryo. They grow when injected to be part of the brain, or part of the skin; it does not matter to which part of the new blastocyst they are added. The versatility of such young cultured cells is referred to by biologists as totipotency. Totipotent cells have a great potential for growth, yet this potential is lost once they specialize to become a specific tissue. Embryonic stem cells are totipotent. They are, like HeLa cells and fertile eggs, immortal—as long as they are prevented from differentiation and development.

Sexual populations of paramecia and other ciliates cannot survive indefinitely without engaging in meiotic sex. The sexual partner may be a complementary gender or even a second haploid nucleus in exactly the same individual cell. Sex itself, not the fact of two parents, is what is crucial. Sex in ciliates provides neither reproduction nor biparentality but a resetting of the clock of the aging process. If no mate is available for their micronuclei mating ritual, a paramecium—doomed otherwise to age and die—undergoes meiosis anyway. Identical micronuclei, after meiotic nuclear division, proceed to mate and fertilize themselves.

Such self-mating is called autogamy. Autogamy comes about when two identical haploid nuclei fuse to form one new diploid nucleus in which the two copies are now identical. As noted, far from increasing genetic variation, autogamy vastly reduces it. One reason given for the advantage of sex is that the increased genetic variation will provide genes from one parent that will, in diploidy, mask deleterious mutations in the same gene from the other. Yet autogamy reduces to zero the chances of masking deleterious genes. Heterozygosity, the state of having two different parental copies of the same gene in a diploid, is compromised when nuclei of autogamous paramecia fuse. Both copies in the diploid are identical since they come from the exact same parent nucleus! Autogamy should lead to unmasked genes and debilitation, yet in paramecia autogamy rejuvenates. Again another standard reason given for the success of biparental sexuality—that it is preserved because of its superior gene masking abilities—is shown to be wrong by the paramecium counterexample.

Self-mating is just as effective in rejuvenation as is sexual mating with others. The formerly dying cells are equally rejuvenated by autogamy as by two-party sex. They resume their rapid pace of growth and are rescued on the road to death. Again, autogamy shows that the true importance of sex cannot be to generate genetic variation. The real point of meiotic sex is to forestall aging, to reset the clock, to repair DNA and organize genes and proteins so that organisms develop and the differentiation of cells and bodies ensues. Together, meiosis and fertilization return nucleated beings to a "zero point" from which they can begin anew their life cycle.

The link between the microbial and animal worlds can be seen in the provocative similarity between microbes with sexual and sexless modes and certain animals which alternate between sexual and sexless reproduction. These animals partake in what scientists call cyclical parthenogenesis: a back-and-forth between sex-requiring and sex-independent modes of reproduction. Consider, for example, a fly, the midge with the species name *Miastor metraloas*. These insects reproduce without benefit of sex in their larval stage.[3] Midges forego adulthood and "normal" sexuality as long as conditions are favorable. But when the environment becomes difficult sexuality returns. Likewise, when certain, plant-eating aphids experience the onset of winter cold, they forego asexual (single-parent) reproduction and begin again to reproduce sexually (biparentally).

The animal phylum Rotifera, with its hundreds of species, exemplifies this same tendency of two-parent animal sexuality to yield to parthenogenetic reproduction on a grand scale. The marine rotifers represent an answer to the female liberation movement because, in all 350 species, not a single male has been espied. Females bear eggs and young that are virgin females. The fresh-water taxa, however, are different. Occasionally, cued by environmental factors, they produce males. Some populations, when they suffer extreme environmental fluctuation, especially the threat of winter, form eggs that hatch males two generations later. Cold water induces these entirely female populations to produce not sons but daughters that bear eggs that hatch into grandsons. Occasional fusion sex in these small translucent water animals coupled with the general absence of biparental sex is correlated with differentiation of bodies associated with certain cyclical environmental changes. Sex is not directly correlated with genetic variation or biodiversity. Rather, harsh environments trigger the sexual response as they recall the imperative of ancestral survival by sexual indulgence.

The origins of death and meiotic sex are traceable to ancient protoctists. *Tetrahymena* and *Paramecium* cells rid themselves of their old macronuclei after mating in a way that augurs the decay after sexual fusion of animal, plant (e.g., flowers) and certain fungal (mushroom) bodies. In these swimmers, the apoptotic death of the used macronucleus follows sexual union and precedes the healthy differentiation of the new macronucleus. Animals, plants and fungi all evolved from such sexual protoctist populations. The devilish genetic tendency for programmed aging and death seems hardwired into meiotically sexual organisms. But this hardwiring, consisting of programmed cell death at least in the form of apoptosis, evolved first in single-celled protoctists similar to ancestors of today's *Paramecium* and *Tetrahymena*. Later, sex, at the cellular level, so intertwined between development and death, was retained by large protoctists and their descendants.

Many species of protoctists which alternate between sexual encounters that decrease the number of individuals and asexual growth that increases the number, consist of closely packed cells with a strong "memory" of acting as individuals. Once populations became new kinds of individuals not only by diversification but by integration, such new, enlarged and integrated organisms were, of course, exposed to severe environmental selection pressures of all kinds. Seasonal environmental extremes shaped their destinies, as always. But new kinds of selection pressures also appeared. Now complex and relatively large sexual multicellular organisms had to battle in a new realm—that of their own growing, maturing and differing populations of aging beings.

CONNECT OR DIE | CONSIDER A RECENT war story from the former Yugoslavia in which a displaced family that had lost the father were charitably given a home. One day the mother and her two children took a walk by the local lake. All three drowned themselves in it. Such group suicide may seem irrelevant to the programmed death of paramecia micronuclei or of the macrophage that induces radiation-damaged body cells to destroy their own DNA and die. People have free will, goes conventional wisdom, whereas cells are automatons. But all humans are made of cells and the body is no mere machine. The human body, like that of all mammals, is better described as a huge integrated community of cells acting under not just genetic, but thermodynamic and other environmental constraints. Cells of the nervous and immune systems die if they fail to establish connections.

So, too, people left out of a "loop"—who feel unloved, have no family, friends or job to "place" them—are more likely to commit suicide, to be depressed or die of "natural causes." Unmarried men die sooner than their married counterparts; people responsible for the care of pets also live longer. We suspect a connection. Perhaps all organisms, from populations of protoctists to people in war-torn Bosnia, tend to self-destruct when removed for too long from interactions with members of their own and other species with whom, historically, they bonded. So-called density-dependent mechanisms of self-destruction appear in animals that over-crowd. To us, programmed cell death is an ancient, well-honed and genet-ically fixed version of a more general and loose tendency for environmen-tally threatened organisms to destroy themselves when their surroundings no longer have a place or function for them.

A description by Craig Holdrege, an erudite high school biology teacher living in Vermont, in an important new book on the genetic manip-ulation of life, is relevant here.[4] In the United States, during the 1940s, Ken Spitz observed children placed in an orphanage at birth who were nursed by either their mothers or by wetnurses. The mothers then departed. According to Spitz, the food, hygiene and medical treatment were among the best he had observed. Because each nurse, however, was charged with caring for as many as eight to twelve children at a time, there was nearly no contact between each infant and the overworked nurses. The time between tending to each child lengthened. Spitz noted, upon his arrival at the orphanage, that not one child had a toy. Moreover, to quiet the babies, the exhausted nurses draped blankets over the edges of the cribs, blocking the children's view of all but the ceiling. After three months of their moth-er's absence, the children started showing signs of deterioration. They cried, screamed, developed insomnia and began to lose weight. At first they had been difficult to comfort; now they withdrew from physical contact when it was offered. They lay on their stomachs in the crib and moved more slowly. They became more susceptible to infectious diseases. Their facial expressions grew rigid and their once vigorous crying turned to whimpering. "After three months the children were completely passive, now lying on their backs. They couldn't turn over. Their gaze was empty and their eye coordination deteriorated."

Of the ninety-one children Spitz observed, twenty-seven were dead at the end of the first year.[5] These babies clearly did not "choose" suicide in the manner we might attribute to an older person. Their decline in interest in the business of living was somehow triggered by a lack of a proper human context. The baby death came autonomically. We could even call the deaths "programmed" if saying so did not show up the obvious inadequacy of this word. It was not so much what was done as what was not done to connect the babies to the rest of human life.

Spitz also observed another home for displaced children, one where similar symptoms persisted but in milder form. The great difference was that in this second orphanage, the mothers returned on a regular basis. When they did, the children recovered rather rapidly. The dying children of the first, tragic orphanage are reminiscent of cells that have not connected. It is as if, in our path-dependent lives, we need to recognize stops on the way, or we will not get from there to here. Genes are a crucial part of the story, but only part. Face-to-face interaction, touching, instruction, imitation, well-timed nutrition and a thousand other daily acts generate health and development. Contact between mother and infant even appears to trigger the circulation of small "mood" compounds—short, specific amino acid chains called polypeptides. These biochemicals, necessary for healthy growth, require not only the presence of genes but the continued presence of maternal attention. For primates like us, early and consistent interaction with the mother is a continuation of the pattern of cell growth, interaction and programmed death that begins inside but continues outside the womb.

Genes are components of cells, and cells are components of organisms. Organisms are parts of populations which are always organized into natural communities of more than a single species. We act, not according to any fixed plan or genetically determined "blueprint," but out of a history of connectedness, involvement and participation.

Perhaps romantic suicide, the self-imposed death of the spurned or unrequited lover, is another example of the great tendency for cells and organisms to destroy themselves when their populations no longer have a function for them. The potential suicide victim, in this case, is deprived of a reason to live not out of any lack of need being met in the present but rather because he has given up hope of his genes establishing their favored contact in the next generation.

We subscribe to L.R. Cleveland's claim that the fusion aspect of meiotic sex originated as desperation in protoctists capable of welcoming each other into their fusible membranes. Mates were at first potential food. No male or female genders existed at first. When starving or thirsty cells fused, it was a haphazard, not an established event. The merging of single cells to make double monsters first occurred not as the result of sexual desire, but of stress. Protoctist relatives, as we have seen, probably attempted to eat the other, imbibing their cohabitants through the feeding process known as phagocytosis. The eater did not always digest the resisting eaten relative, however. In crucial cases eater and eaten merged forming a single larger cell. Doubled monsters became the first diploids, with their doubled sets of chromosomes. A slight adjustment in the timetable of cell reproduction—delay in the reproduction of the spindle-attachment body, the centromere, relative to chromosomal DNA—would have provided relief of diploidy.[6] This natural delay returned fused cells to their original, single-cell state. Under certain cyclical, seasonal conditions, monstrous doubling and its return to sleek normals, as we saw, was positively selected. Like electricity and automobiles in modern human life, yesterday's luxury becomes the necessity of today. Over time, feeding frenzies and protoctist fusions became fertilization and reproduction. Long ago, the pleasures of gastronomy and eroticism were one and the same, and gender did not exist. As long as the cannibals with indigestion survived the fusion, they became new individuals, merged beings, the result of a desperate union. Reductive cell division (meiosis) and growth (multiple mitoses) replete with programmed death followed. From such an odd concatenation of circumstances, our gendered bodies have come.

Our lives are circumscribed by sexuality. We come from a void, we go to a void. In between we live. Our lives, and our deaths, are mediated by sex. We appear on this Earth through an act of sex, which begins an irreversible act of aging. Our parents embraced sexually at least once. We were born and, someday, whether or not we participate in sexual coupling, we will die. As a form of material organization, we are over three billion years old, but as conscious personalities we are at most a few decades.

Sex cyclically produces bodies from fused sex cell through embryo formation, childhood and adolescence. But this individual form of material organization is limited. Each of us inevitably ages and dies. The only way we preserve the dissipative process of our human body is to engage in fertile sex to reproduce; sex both inaugurates and seals our fate as individuals. The Faustian price we pay for the sex act that initiates the complex tissue differentiation of our offspring, including the clump of neural tissue called the brain, is encroaching death. Our imperative is to reproduce, within the time span of our fertility, a body as a particular form of material organization away from thermodynamic equilibrium.

The Judeo-Christian association of sex with "the Fall" is scientifically resonant. When meiotic-fertilization sex cycles evolved in our protoctist ancestors, these communal descendants of immortal bacteria lost their innocence. Both religiously and scientifically, sex has spelled death ever since. The evolution of bodies begun each generation by sex brought an end to continuous energy degradation by cells. It planted upon the lips of early life the kiss of cyclically recurring, sexually mediated death. The result was the discontinuity of multicellular bodies. The "kiss of death," with its implicit suggestion of a connection between sexual reproduction and the inevitable demise of sexual bodies, is an apt metaphor for the evolution, several hundred million years ago, of discardable bodies that serve as temporary vehicles for the so-far-immortal, gene-containing sex cells.

five

STRANGE ATTRACTIONS: SEX AND PERCEPTION

The final cause of this contest among the males seems to be, that
the strongest and most active animal should propagate the
species, which should thence become improved....
[I]f vegetables could only have been produced by buds and bulbs
and not by sexual generation...there would not at this time
have existed one thousandth part of their
present number of species.

— Erasmus Darwin

When one is in love one begins by deceiving oneself.
And ends by deceiving others.
That is what the world calls a romance.

— Oscar Wilde

IN THE WAKE OF SEX AND DEATH | AS WE HAVE SEEN, fertilization sex evolved because our protoctist ancestors survived the seasons. Sex itself was not selected for in animals. Rather sex was the only way animals could develop. Nonetheless, in the big picture of planetary evolution, sexual species, once they evolved, proliferated exuberantly. Moreover, populations of sexual animals and plants enacted life's thermodynamic purpose. As we will see, they more quickly and more extensively than their microbial predecessors broke down the solar gradient. On a smaller scale, any gendered animal is attracted to certain traits in a potential mate and breeds only with those who possess them. As such, animals tend to reproduce sexually selected traits in their offspring. Such traits may range from the showy oecelli ("eyes") of the peacock's tail feathers to the blue scrotum of the mandrill. [PLATES 35 and 36] In this chapter we first explore the wider context of meiotic sex, then move on to traits, including those of our own species, which have arisen in the wake of sexual reproduction.

The biparentality of animals produces blastulas from fertile eggs. Fusion sex, in its cellular core in all major groups (phyla) of animals, is a dependent path that may well be impossible to eliminate. (see **Appendix**.)

Sexual species, both animal and plants, which have dominated life for the last half a billion years, seem to be earning their keep on a global scale. Just look at the vast number of surprising and brilliantly colored species in the equatorial rain forest [PLATE 37] for an index of sexual species' great biodiversity. Variation in genitals or flowers, two direct signs of sexuality, are inevitable byproducts of any given species' path-dependent imperative for sexual reproduction. We suspect that the sheer variety of plant, animal and fungal sexuality, where each individual belongs to a different population of potential mates, increases the rate of energy degradation and entropy production. Ecosystems rich in sexual species, such as the Amazon rain forest, more efficiently reduce the solar gradient than the less sexual bacterial ecosystems beneath the ice of Antarctica. Measured from space in the infrared portion of the electromagnetic spectrum, communities such as those of Amazonia more effectively cool and produce entropy than do those of their less varied cousins.[1] While sex is not directly selected for in either plants and animals, sexual species abound because specific plants and animals are selected. These sexual species, especially trees, are key to Earth's most complex and greatest entropy-producing ecosystems.

Life is sentient. Life affects its own evolution. Even seemingly simple choices made by bacteria swimming toward food lead, over time, to new evolutionary directions. A bacterium that swims toward its food which is the waste of a second type of bacterium, for example, will likely continue its association. Patchiness of habitat occurs as one type of bacterium gobbles the waste from a second and stable associations form. Swimming protoctists, required to survive winter by fusion sex, associate not only with their food but also with their mates. With the evolution of mandatory cyclical fusion sex, protoctists and animals on the move became actively involved in their own evolution. Charles Darwin, Erasmus's grandson, pointed out two ways in which sex accelerates evolution, both of which, in contrast to natural selection, he called "sexual selection." Female animals were in some cases, he noted, in short supply relative to the number of males that might breed with them. Charles Darwin described both competition between males, leading to stronger, more fit males, and female choice, in which females select the males with whom they breed. He noted that animals exercising choice can influence traits of the next generation.

In the days since Charles Darwin, another form of sexual selection, sperm competition, has come under the scrutiny of scientific researchers. The victors in sperm competition are not those able to displace the bodies of other males by fighting, but those who are able to displace competitors' sperm by means such as producing greater numbers of sperm or ejaculating further. One common strategy that evolved independently in spiders,

PLATES 35-36
Mandrill face and rear, showing similar
pattern of coloration front and back. "(N)o
case interested and perplexed me so much
as the brightly-colored hinder ends and
adjoining parts of certain monkeys.... It seems
to me ... probable that the bright colors,
whether on the face or hinder end, or, as in
the mandrill, on both, serve as a sexual
ornament and attraction." – Charles Darwin
[*Toni Angermayer and Tim Davis/Photo Researchers*]

insects and mammals is the formation of extremely sticky semen that effectively blocks entrance to the sperm of subsequent suitors. Certain male rodents even synthesize "vaginal plugs" which they deposit with their penises. In animals, especially notable in social primates like ourselves, the ability to vary behavior in search of sexiness, to perceive subtle distinctions in the shape, size and color of bodies and to deceive ourselves and others, has become incorporated into the evolutionary process.

"Do you see no further than this façade—this smooth and tolerant manner of me?" asks Walt Whitman in his famous work, *Leaves of Grass*. But, as organisms, "façades" are often the only clues we have. No animal directly perceives genotype, that is DNA making up the genes in the chromosomes of a prospective mate. She or he only notes the phenotype, or how those genes are expressed to make the whole animal. Indeed, any perspective mate only perceives the outside of the other's body, from a certain angle, in a certain light. The poets speak of the special light in which the beloved is bathed. Novelists write of infatuation. Many romances depict the dangers lovers bring upon themselves by following their urges to sexually couple. Powerful biochemical changes—from the production of the natural amphetamine-like drug phenylethylamine during the initial "rush" of physical attraction to elevated levels of the hormone oxytocin during and after orgasm—correlate with lust, love and pair bonding. These inevitable body-produced, mind-altering drugs are part of the biochemical arsenal by which our natures entice us to seek mates and produce offspring. We, like many other animals, sometimes risk our own survival for the chance to inject our genes into the next generation.

But since we are imperfect beings, not omnipotent gods, our perception of desirable mates evolved from the same apparatus used by our unsexed ancestors to locate food. As a result, we can, and do, send out and respond to false signals regarding the sexual desirability of others. We are quite easily duped, deluded, enticed and enthralled. Detailed in this

---

PLATE 37
Rainforest sexuality.
[*Alexis Rockman*]

chapter is an important part of the evolution of sex: how perception mandatory for mate coupling and offspring production continues to shape the body and mind. The evolutionary precursors to human gender recognition, human fusion, sex and love are explored. Lust and love are far more crucial to evolutionary continuity than they are to the well nourished individual animal. They entice us to sacrifice our individual bodies, destined, as we have seen, to die anyway, for the sake of future sexually produced beings. We are ephemeral. Sex remains.

Deception is based upon perception, and sexual deception is based upon sex and perception. How did sex in mammals evolve? Is mating mandatory? How, amid the plethora of stimuli, do we locate a suitable mate? How is our sexuality related to our perception of other life forms? We compare perception, deception and mating strategies across a range of species in an effort to understand the source of human sexual behavior.

Our perspective has been clear. Life itself comes from the thermodynamic unfolding of the physical universe. Meiotic (fusion) sex is a biological phenomenon that evolved in stressed protoctists. Our animal ancestors which emerged from bizarre protoctists could not develop embryos or adults without sperm penetration of the egg. Sex, in us, is not dispensable; development and reproduction of all members of the 38 phyla of animals require meiosis and fertilization. (See **Appendix**.) We believe our scientific analysis of these path-dependent processes helps us understand even subtle aspects of the evolution of sexuality such as how animal perception, over evolutionary time, shapes animal bodies.

PERCEPTION, DECEPTION, AESTHETICS | Perception, deception and aesthetic sensibility have been evolving in animals during the more than 541 million years of their existence on Earth. No wonder, then, that these very capabilities have both been honed by, and themselves influenced by, animal evolution. Despite our arrogance and our human-centeredness, the results of many experiments and observations force us to realize that the processes of thinking and feeling are not solely limited to humans. Perception and action stimulated by perception are properties of life—even bacterial life. Indeed, all of living nature is an interacting, perceiving phenomenon.[2]

A few types of swimming bacteria, tolerant of but not dependent upon atmospheric oxygen, are magnetotactic. They respond to magnetic forces and swim—in the northern hemisphere, toward the North Pole, and in the southern hemisphere toward the South Pole. These living compasses harbor tiny crystals of magnetite aligned in rows within their bodies. The magnetite crystals help them orient themselves along Earth's magnetic lines of force. These bacteria end up not at Earth's poles but safely in the food-rich, oxygen-poor sediment in which they thrive. Even the smallest non-magnetic bacteria perceive, and since they can locomote we can say that they behave. Tiny bacilli or longer filaments swim to align themselves along a gradient of increased sugar concentration as they approach a perceived food source. Sensing, they act on the basis of their sensations. Many bacteria, and their hypersexual protoctist descendants, move toward or away from light or oxygen. The perceptual ability of life is directly related to its gradient-reducing activities. Living beings identify gradients—nitrogen, sugars, acids, light, heat and "loneliness" (i.e., separation from other cells or organisms)—and respond accordingly. Then, they turn perceptions into action and reduce these gradients.

Protoctists' "survival through sex" later evolved into sex as a requirement for reproduction. Many of the perceptual abilities that had evolved in protoctists for food acquisition, enemy avoidance and habitat location were later modified and reused by the ancestors of animals. Given that protoctists—long before they were sexual—were bacteriovores, protoctistivores and even cannibals, we suspect that the joys of eating evolved long before those of mating. From the beginning, all mobile forms of life had to

procure food and shelter. Sexual protoctists and their animal descendants also needed mates. For hundreds of millions of years, the reproduction imperative forced sexual protoctists and their early animal descendants to distinguish potential mates among a world of distractions.

The beautiful, beheld in the biased eyes of members of his or her own species, survived. Beauty may be skin deep and fashion a synonym for the superficial, but over the eons, among those forced to sexually reproduce, sexual attraction has been a matter of life, death and continuity. False or not, attractiveness is and has been reality. Inferring the genetic desirability of a potential mate is of the utmost importance in all sexually reproducing beings. "Bright eyes and bushy tails" in mammals are clues to fertility and capacity for effective parenthood. Skin texture, finish or gloss of fur coat, and peak breeding age, for example, help to determine the fertility of a potential mate.

SEXUAL SELECTION AND FEMALE CHOICE | PONDERING THE PEACOCK, Charles Darwin was the first to suggest that animals have certain distinctive features, not because those features themselves necessarily have survival value, but because, when the other gender finds them attractive, they inspire fertile matings. [**PLATE 38**] Assortative mating refers to fertile unions of similar-appearing organisms. Assortative mating leads to distinctive populations which share common traits. Teenagers who primp and preen for their peers provide examples of how new mating tastes develop by manipulation of one's appearance to appeal to the other gender. We all know of subgroups of populations that repeat customs that lead to provisional assortative mating communities. Traits of intrabreeding subpopulations—human examples might include nose rings and lipstretchers, ducktail hairdos and stilts—appear quirky or unattractive to some members of the same species. All of us are sensitive to tentative reproductive isolation seen in humans as the insularity or snobbishness of cliques. Sexual passion runs deep. Indeed,

PLATE 38
The splendid display of the peacock.
[Frans Lanting / Minden Pictures]

PLATE 39
**Jungle fever: a praying mantis mounting a
chipmunk.**
[*Alexis Rockman*]

unfilled sexual urges may be so strong that, should organisms lack an appropriate partner of the same species, they sometimes mate—unsuccessfully from an evolutionary point of view—with members of other species. [PLATE 39]

In *The Descent of Man, and Selection in Relation to Sex*, Darwin proposed that features such as the peacock's tail,

> ...serve only to give one male an advantage over another, for the less well-endowed males, if time were allowed them, would succeed in pairing with the females; and they would in all other respects, judging from the structure of the female, be equally well adapted for their ordinary habits of life....for the males have acquired their present structure, not from being better fitted to survive in the struggle for existence, but from having gained an advantage over other males, and from having transmitted this advantage to their male offspring alone. It was the importance of this distinction which led me to designate this form of selection as sexual selection.[3]

Sexual selection leads organisms to influence their own evolution. Darwin was steadfast in his insistence that females, even insect females seemingly driven by blind instinct, can dramatically impact their own futures by choosing to mate with only certain males. For Darwin, it was an "astonishing fact that the females of many birds and some mammals...and...even more astonishing...reptiles, fish, and insects" exercise "female choice." Although originally disparaged, even by those evolutionists who endorsed natural selection, Darwin's sexual selection has since been vindicated.

In 1982, Scandinavian zoologist Malte Andersson, an expert on sexual selection, published reports on the effect of clipping some 15 inches off the tails of male long-tailed widowbirds (*Euplectes progne*) which he glued onto other males.[4] The super-tailed males became more successful fathers, as measured by the greater number of new eggs or chick-filled nests in their territory. Captive zebra finch females (*Poephila guttata*) prefer males with legs ringed in red and orange or green loops. The male zebra finch spurns black-ringed females, preferring to inseminate those females ankleted in blue or orange, even though the black-ringed females enjoy a reproductive advantage and produce more healthy chicks. University of California biologist Nancy Burley, who uncovered these surprising avian aesthetics, also discovered that females preferred to mate with cocks wearing certain colors of hats.

TAIL FEATHERS AND BEYOND | TELEVISION AND MOVIES reinforce norms of sexual attractiveness on an increasingly global scale. Consider the slim, tan supermodel. Our current feminine standard of beauty differs greatly from the Renaissance ideal of healthy plumpness, as found in the paintings of Botticelli and Rubens. Today, we muse at the Victorian female ideal of pale, frail, even tubercular femininity. Change in fashion toward skinnier females may or may not reflect survival value. Perhaps, in times of relatively limited resources, the bodies of wide-hipped, plump women indicate a potentially greater ability to bear and nurse infants. Alternately, in modern, more densely populated urban cultures, slimness may indicate women who are easier and less expensive to support. Even in the limited time frame of modern Western humanity, drastic changes in fashion have occurred, with potential evolutionary consequences. Although diametrically opposed, both slim supermodel and robust Renaissance beauty are examples of features favored by sexual selection.

Some use the term runaway selection to refer to a type of natural selection that is based on no obvious survival advantage, other than the peculiar, reproduction-reinforced attractiveness of the possession of a certain trait by sexually desirable members of the opposite gender. The desired traits may even be detrimental to their bearers. In the handicap principle of Israeli zoologist, Amotz Zahavi, the specific pattern of stripes on a zebra, the oecelli of a peacock and the ring of pigment at the base of an animal's long neck, signal specific show-off but self-handicapping kinds of messages. "An animal with a long neck may display the length of it by having a handicapping ring around the neck. Individuals with short necks will look even shorter necked: 'My neck is so long I can even afford to make it look short.'" His theory suggests some whimsical features, for example, the brilliant reds of the male hummingbird, *Spathura underwoodi* who also sports long, ornamental tail feathers, may advertise quirkily evolved bravado. The allure may even be due to the lack of any survival value to the traits in question. The boy, jumping on the sidewalk, does twenty fast push-ups to impress the girl. This behavior wastes energy. Even if it slightly handicaps the boy who performs, it also signals an available mate—a worthy male who has energy to spare.

But the more spectacular examples of alleged reproduction-reinforced attractiveness are structural and physiological, rather than behavioral. Male pelicans (*Pelecanus onocrotalus*) grow large bumps on their beaks during mating season. These protrusions are apparently highly attractive to females, even though they interfere with vision and the catching of fish. The non-survival trait of slender women even begins to makes sense, using Zahavi's model of female handicapping. It is as if they are saying "my body is so desirable as a baby maker, I can afford to make it look like it couldn't even support a baby."

Zahavi's handicap principle is an example of how features are explained by reference to their survival value. Although Darwin distinguished sexual selection from natural selection, ultimately the former is an example of the latter. Even if an animal is vigorous and in perfect health, his genes fail to enter the next generation—that is, they are selected against—if he is infertile or if no one mates with him.

Darwin's distinction between sexual selection and natural selection implied the two were independent processes. His countryman and co-discoverer of natural selection, Alfred Wallace, downplayed sexual selection as an evolutionary force. Wallace did not believe that the bright colors of male birds were chosen by females, suggesting instead that spectral flashiness was the natural avian condition. In other words, males had not evolved brightly colored feathers, rather females, because of nest predation, had evolved drabness as a form of camouflage.

Not believing that animals choose mates sporting brilliant colors, Wallace thought that natural selection (which like Darwin he distinguished from sexual selection) obscured "brilliantly metallic blues and greens...the most splendid iridescent hues."[5] Red blood and white bones are colors resulting from the iron and calcium composition of these respective body parts; they are not adaptations, nor were they selected for. Wallace rejected the idea that mate choices played any significant role in bird color or blood color evolution.

Prominent biologists, including Julian Huxley (1887-1975), joined Wallace in doubting Darwin's theory that mate choices influence the course of evolution. Huxley, Wallace and others failed to see how nonhuman females exercising mate choices over long periods of time could lead to significant body changes. Exquisite plumage, such as that of the male Argus pheasant, which, Darwin wrote, was "more like a work of art than of nature" did not, in their opinion, come from female pheasant whim. Wallace put it this way: "I do not see how the constant minute variations, which are sufficient for natural selection to work with, could be sexually selected....How can we imagine that an inch in the tail of the peacock, or ¼-inch in that of the Bird of Paradise, would be noticed and preferred by the female?"[6] Ostentatious traits might be an adaptation for mate recognition, a result of higher male metabolic rates or an advertisement to predators that males were not too tasty. That these traits reflected female choice by "lower" animals seemed impossible.

Consider again Andersson's experiments involving the elongation of the tails of long-tailed widowbird males. Wallace would have emphasized the possibility that longer tails are only superficially superficial: they may, for example, be correlated with the success of males warding off parasites, presumably more visible on longer tails. In recent years the argument has been explicit. According to Hamilton/Zuk theory, conspicuous ornamentation attracts females because it indicates hereditary resistance to parasites. Thus, biologists William Hamilton and Marlene Zuk, like A. R. Wallace, reject the idea that female animal choice is frivolous. What appears to be frivolous display in the form of bright colors and long tails may be symbols for judging parasite-resistant genes. In 109 passerine (song) bird species Hamilton and Zuk found chronically blood infected birds tended to lack the bright colors and complex songs of their healthy brethren.[7]

DAMSELFLY PENISES AND ORANGUTAN LOVE | Taxonomists classify damselfly species by their extremely varied male genitalia. These genitalia are almost baroque in their inclusion of flanges, barbs and scoops that remove previous male's sperm. [**PLATE 40**] Many female insects have clasping apparatus associated with the male genitalia. Some female insects even have complementary indentations that snugly clasp the protrusions of males. A theory of the evolution of elaborate male insect damselfly genitalia attributes them to a "mechanical conflict of interest." Males develop more effective insemination equipment to overcome female resistance. But, as entomologist William G. Eberhard points out, a true, all-out battle of the sexes would suggest "concomitant changes in female and male genitalia, and anticlasper structures in females."[8]

Many examples exist to the contrary. These include female Cincidellid beetles, which have indentations to receive the male organ attachments in the grasped area, and the males of the dragonfly species (*Epigomphus quadracies*) which have abdominal appendages that fit into sockets on the heads of females. Although, during the time that care of offspring is paramount, it may be to the female's advantage to reject all male advances, usually the female's genetic interests are best served by rejecting only some males—the

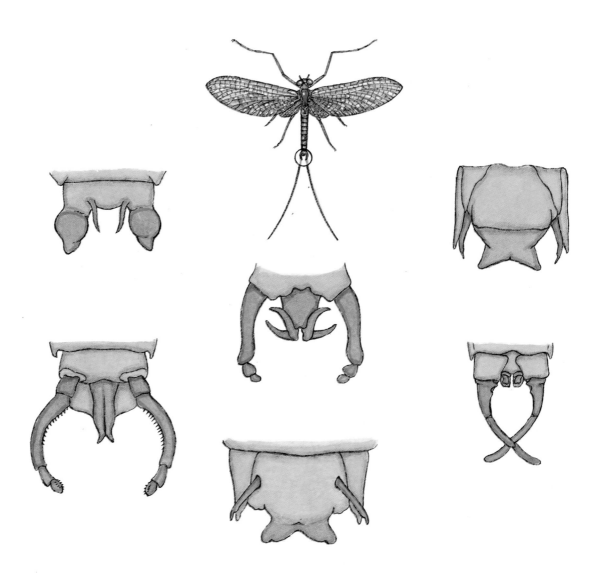

PLATE 40
**Damsel fly genitalia. Blue: claspers;
pink: penile.**
[*Patricia Wynne*]

less adept breeders. As metaphor, the "battle of the sexes" is as overreaching as the "harmony of nature" or the "fragility of the planet." At best "the battle" is only part of the story.

One explanation for male genital diversity is that it is established and maintained by female choice (females choose males with the most effective sperm delivery apparatus). [PLATE 41] The evolution of mating apparatus is complicated by the fact that females do not simply choose males, but instead, choose from among males who are themselves competing for females. Male insects compete with each other to inseminate females. Bigger bodies, faster sperm and more aggressive sexual behavior may all bestow advantages to the evolution of males possessing such traits. Males with genitals more effective in grasping, entering or stimulating female reproductive organs, sire more offspring—thus they enjoy a persistent evolutionary advantage.

Females who choose the inseminating males propagate their own genes in their sons. Female choice is restricted, since any who fails to choose the male with effective sperm implantation equipment may lose out in the long run to she who does. Like a spectator "choosing" a card forced by a magician, this is a Hobson's choice.[9] Over evolutionary time and given male-to-male competition, traits leading to high male fertility tend to replace those leading to low fertility. Females, therefore, develop increasingly acute means of detecting the best-breeding males. By recognizing superior male sperm-delivery equipment by visual or tactile cues, her genes tend to propagate through the population. The impetus to evolve aesthetic senses connecting sexuality to the rest of the nervous system occurred as neurological pleasure inspired females to breed with titillating, fertile males. The capacity for female orgasm reflects unconscious neurological connections that provide an index of male sperm implantation equipment.

The runaway diversity of damselfly penises comes from damselfly spe-

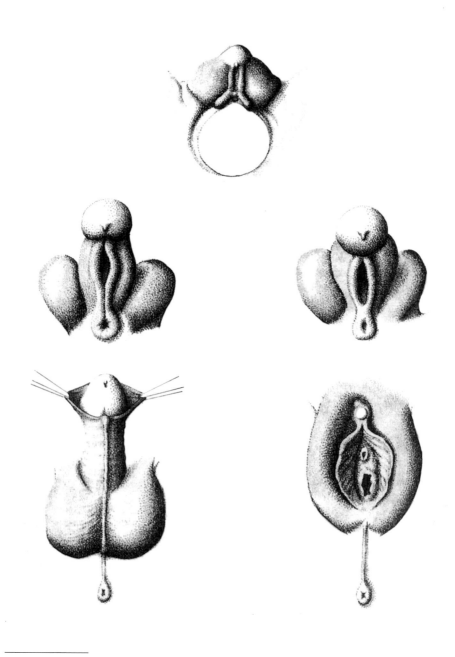

PLATE 41

Human males are indistinguishable from females at first, in *utero*: both have mounds of tissue where their genitals will later appear (top). The tip elongates, becomes the penis with its foreskin, and the sides fuse along the midline and swell to become the testicles in the male (left), whereas in the female the tip differentiates to form the clitoris and the sides become the vaginal labia (right).

[*Christie Lyons*]

ciation. Any individual damselfly who fails to identify a potential mate of its own species leaves no offspring. Males compete to fertilize females who, in turn, compete to recognize and couple with the most fertile males. Male genitalia, like other male characteristics, are chosen, consciously or unconsciously, by the egg-making female. "A female," as Eberhard writes, "could discriminate among males' genitalia on the basis of their ability to fit mechanically into her, or through other sensations occurring in her genital region. Once such female discrimination was established, selection would favor any male that was better able to meet the females' criteria (by squeezing her harder, touching her over a wider area, rubbing her more often, and so on), even though his genitalia were no better at delivering sperm than those of other males. The ability to `convince' the female enhances male fitness."[10]

Part of the confusion surrounding sexual selection in the evolutionary literature has to do with the lack of a coherent philosophical theory of what constitutes a sign. In a universe filled with complex behavioral communication, language naturally evolves. One single purpose or definitive meaning seems unlikely to be retrieved from the entanglement of mixed messages in a world full of signalers. Hats, for example, are of course functional—they protect the head from heat, rain, and sunlight—but they may also have sexual value. Although we did not evolve with hats on our heads, these body extensions, just like the swollen beaks of male pelicans, affect the intricacies of sexual attraction and mate choice.

Novelist Milan Kundera calls the hat a magical object: hats, which extend the head and frame the face, create an aesthetic aura which cannot be reduced to mere function. Some body extensions have sexual significance in that they attract. A female pelican likely shares this opinion with her half-blind potential mate. An extraterrestrial "Wallacean," seeing hats as technological extensions of the human body, might explain them as cranial thermo-regulators, desirable to mates because they increase the hardiness of their wearers. Hats, some of them, attract potential mates because to them they "look good." Other headdresses may repel, for example, by signifying the approach of a violent tribal enemy. Hats have no absolute meaning, but they do signify. Significant differences in bodies and body extensions emerge from selective breeding histories over vast periods of time and lead to splintered populations. Eventually, bearers of specialized features, whether sexual or non-sexual in origin, no longer breed with those not bearing them.

Technology, culture, civilization and art do not distance us from our animal selves but, instead, accentuate and expand our animal natures. Our mammalian ancestors were four-legged vertebrates whose principle means of ammunition and defense were their jaws. Perhaps the beauty we see in high cheekbones reflects an ancient vertebrate awe of maxillary power. As hierarchical mammals impressed by large size, people cheer the booming voices of leaders poised on elevated platforms. Adolescent youths prefer oversized clothes that make them look larger or tight clothes that make the well-outlined wearer seem smaller, more angular and proportionally more well-endowed. Despite our pretensions of superseding our "primitive" animal ancestors, we remain tailless, jealous primates fond of sexual display. We inevitably respond with sudden awareness to the exposed pubis, penis, buttock or nipple. We attend predictably to a changing variety of sexual signals. Calf inserts, codpieces, miniskirts, trendy sunglasses, epaulets, wigs, lipstick, necklaces and rings show different degrees of functionality, but all communicate sexually. Fashion accentuates the ancient connections between physiological variation and sexual success. Our reactions to perceived differences among mateable members of our species are far from arbitrary: they are profound responses to millions of years of evolution.

UNCONSCIOUS INFERENCE AND TRUST | INHERITED BEHAVIORAL tendencies set up animals to be fooled. Take gullible gulls; if they glimpse a different species at a crucial moment in their development, they will later become attracted to members of the "wrong" species. With evolutionarily futile infertility, they attempt to mate for the rest of their lives with inappropriate mates. To human observers, herring gulls, Thayer's gulls and glaucous gulls—which all nest together in areas of Canada and Iceland—appear identical. However, the three never interbreed because, as chicks, they imprint on the iris and eye ring color of their nest mates whose natural "make up" ranges from pale yellow to pale orange to pale violet. [PLATE 42] When banded chicks were transplanted to nests of a different color or exposed to painted eye rings, they were fooled.[11] The nestlings grew to attempt mating with members of other gull species who rejected them because of their "wrong" eye color.

Domestic peeping chicks, even those restrained from physical contact, are tended by their mothers. These same chicks, covered with a soundproof glass bell, are ignored by the mother hen, indicating that the mother responds only to the chirping sounds, not the sight of her young. Recent experiments conducted with fruit flies have isolated the genes which control courtship by coding for the development of those areas of the fly brain in charge of mating behavior. Humans are not exceptional, rather our complex perception, coupled with an absolute requirement for heterosexual indulgence as a ticket to enter the next generation, permits us to exquisitely fool each other and ourselves.

The sights and sounds we perceive do not simply represent the passive reception of external phenomena through the transparent windows of the senses, but are constructed actively, intuitively and often incorrectly. Nonetheless, we cannot begin to picture the world without data streaming through our senses. Animals, like microbes before them, have developed economical means for perceiving the surrounding world, particularly those parts of the world important to perpetuation in the evolutionary game of life. An early investigator into how we see the world, Hermann von

Helmholtz (1821-1894), characterized human perception as "unconscious inference."

Limited by our big-brained bodies within a larger universe, we are prone to impulsive conclusions before the evidence is all in. Indeed, since we cannot possibly obtain all pertinent evidence before we make any decision, we, like all other forms of life, always act on the basis of limited information. Nonetheless, under given conditions—open air, natural light, the ability to walk around and handle objects, to view them with two eyes in three dimensions—our mental models of the world appear remarkably reliable. Perception, from the chemical sensation of microbes to the sonar of dolphins, is crucial for survival and genetic continuity. The internal images we and fellow animals generate must correspond somewhat to a palpable reality if we and they are to leave descendants that survive. Yet no internal reconstruction of reality can be exact.[12]

Unconscious inferences can be subtle and pervasive. Consider an orange. Although yellower in the direct sunlight of afternoon, bluer in the indirect rays of sunset and mottled in the shade as it reddens at sundown, we continue to perceive it as the same-colored orange. Cone cells that line the retina of our eyes, combined with neural processing in our brains, lead us to automatically compensate for the color changes. We construct the same consistent view of that same appetizing piece of fruit. However, were the orange viewed under a sodium or other lamp with a different wavelength of light than that to which we were evolutionarily accustomed, the same orange may look very different, even repulsive.

Stanford University research psychologist, Roger Shephard explains unconscious inference in the following way:

> Through the millions of generations of its evolution in the three-dimensional world, the visual system has become highly efficient at providing us with an accurate and reliable internal representation of what is going on in that world. What our experi-

PLATE 42
Imprintable irises and eye rings of the gulls.
[Patricia Wynn]

174

ence gives us is, in a sense, the 'illusion' of direct, unmediated access to the external world. Our perception of a stable, continuous, and enduring three-dimensional surrounding retains no trace of the prodigiously complex neuronal machinery that so swiftly constructs that experience. Nor are we aware of the shifting, intermittent, pointillist, upside-down, curved, two-dimensional patterns of retinal excitation from which the machinery of the brain constructs our visual world.[13]

If aspects of the environment are unchanging or change in predictable ways—as do the relationship between objects and their shadows—animals evolve perceptual shortcuts. Unconsciously inferring, they survive, mate and leave more offspring.

THE PARABOX | EXACTLY HOW THIS PROCESS of unconscious inference occurs is open to discussion, but it is beyond doubt that we are easily fooled. Consider the ancient "moon illusion." Near the horizon, especially in autumn, the Moon looks larger and closer to us than the same Moon viewed overhead a few hours later. An entire academic book, many hundreds of pages long, has been devoted to multiple explanations of this would-be simple phenomenon. "How little we know how little we know," wrote Nobel laureate Max Delbruck.[14]

The Parabox is a three-dimensional optical illusion invented by sleight-of-hand artist Jerry Andrus of Albany, Oregon. The Parabox may be constructed by photocopying, cutting out and assembling **PLATE 43**. Because one of our perceptual "short-cuts" is to mentally construct volumes from the appearance of surfaces, the inverted Parabox, when viewed with one eye at slightly less than arm's length, will seem to jut up and move eerily. We infer a sphere from a circle, a cube from a square and parallelograms. Most objects, unlike the Parabox, do not appear "inside-out." We perceive solid, three-dimensional objects from flat two-dimensional surfaces.

Animals have learned to exploit this tendency to make themselves appear more imposing by presenting convincing surfaces such as triangu-

---

PLATE 43
Parabox—illusion cutout. Photocopy this image, and then fold and paste to create Parabox.
[*Jerry Andrus*]

lated sails, wings and throat fans, as did many of the once fabulously successful giant reptiles. Like the builder of a movie set who saves money by building façades rather than edifices, many species of animals have evolved large-looking surface or display features. Threatened male lizards expose their larger-looking sides to each other in mating rivalries. Appearances and behavior changes always evolve more quickly than do changes in entire bodies. Opening both eyes destroys the three-dimensionality of the Parabox, but the lesson that perception is not perceived but constructed, that it is not passive but involves continuous mental activity, remains. We actively construct our world, especially our perceptions of the other and his or her motives, from incomplete data. The potential for misinterpretation, manipulation and deception is vast.

When we view the Parabox with both eyes, the illusion evaporates, exposing our all-too-convincing mental constructs. Deterministic mathematical precision, while theoretically satisfying, never perfectly describes the messy reality of real life. The rising importance of probability theory (in sciences ranging from population biology to thermodynamics and quantum theory) highlights imprecision, risk and likelihood. Calculations of the probable relative to the exact force us to grudgingly recognize that our hope for the final mathematically correct description of the universe recedes into impossibility. Neither our senses nor our mathematics are perfect even in principle.

LOVE DRUGS | NATURE URGES US to reproduce, not particularly caring if we kill ourselves in the process, as long as we make more of our kind. Drugs, pure chemicals with highly specific and predictable effects, have profoundly influenced sexuality since Archean times when bacteria ruled Earth. The antibiotic tetracycline, for example, when placed in a bath of bacteria, increases the rate of bacterial sex a thousand times, as measured by gene transfer. Simple chemical compounds induce deep physiological, including sexual, changes. Internally and externally derived mood-altering drugs abound.

Some natural drugs are so powerful that they would probably be banned if they were sold for profit rather than inherited as part of our natural perception- and emotion-modulating biochemistry. One such drug, oxytocin, a muscle contractor, and one of them, induces milk flow in new mothers. Oxytocin-blocking compounds, given to post-parturition female rats separated temporarily from their offspring, lead these new rodent mothers to lose all interest in succoring, cuddling and caring for their young. Blood levels of oxytocin in mountain voles, burrowing wild rodents who typically abandon their infants soon after birth, are lower than those of prairie voles, whose mothers are more attentive. A "cuddle chemical" that stimulates contraction of smooth muscles, oxytocin mounts in maternal blood after suckling. Oxytocin levels increase five times in a man's blood during orgasm and are found in even higher concentrations in a woman's blood after sex. During nursing, as cortisol levels and blood pressure drop, the blood vessels in her chest dilate, creating warmth for her suckling infant. Unlike male warriors pumped up on hormones such as cortisol and epinephrine (adrenaline) that increase blood sugar and blood pressure levels, the blood sugar levels of the nursing mother decline.[15] In contrast to the threatening, testosterone-stoked warrior (or his modern equivalent, the steroid-replete athlete), she is calm, nurturing and non-threatening. The oxytocin levels produced by her hypothalamus increase, and she becomes calm.

Clearly, this "hug drug" has played a role in the sexual evolution of human beings. Perhaps the rise of oxytocin levels in both our male and female ancestors made them more likely after sexual intercourse to enjoy each other's company. As our ancestors increased their familial unity, they became brainier, more loving and social. Working together bettered predator defense and food acquisition skills and, as a result, certain early human tribes, bonded together in part by the love drug oxytocin, may have displaced other less closely knit ones.

Oxytocin can be contrasted with phenylethylamine, tentatively identified as the "infatuation drug."[16] Mice, rhesus monkeys, and other mammals injected with phenylethylamine (PEA) moan with pleasure, exhibit courting behavior, and addictively press levers to obtain more of the drug. PEA, a natural amphetamine-like molecule—a natural "speed" for lovers— resembles plant-derived hallucinogenic drugs. [PLATE 44] Two people deep in the throes of mutual attraction have elevated levels of PEA, which accelerates information flow between nerve cells. A kind of natural mood elevator, PEA may be deficient in what Michael Liebowitz and Donald Klein of the New York State Psychiatric Institute call "attraction junkies," those promiscuous adults who seek new sexual partners to feel the initial rush of infatuation before, as the saying goes, "the honeymoon is over." Most couples, while perhaps less excited, do not abandon their relationships after the waning of PEA-mediated infatuation. In fact, diminishing PEA levels in attraction junkies seem to produce outright depression. Liebowitz and Klein administered antidepressants (MAO inhibitors) to attraction junkies and found that they responded much more quickly to therapy than those untreated with antidepressants.

We also know that PEA not only mediates feelings of infatuation but also danger because levels rise during thrill. An imbalance or lack of PEA seems to induce some people to seek out dangerous situations. The "high" of the gambling win may also be due to PEA. Feelings of well-being, pleasure, and heightened sensory awareness, produced by falling in love, may be intensified by exposing oneself to situations of risk and fear. As nature's own speed, PEA confers a sort of alertness and confidence conducive to handling new and dangerous situations. For example, researchers found that men who met women on a wobbling suspension bridge were more likely to request dates than those who met women in the relative safety of a

PLATE 44
Chemical structures showing similarity of phenethylamine (a), a natural "love drug," to (b) amphetamine ("speed").

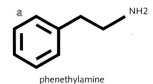

phenethylamine

α-methylphenethylamine

college campus or office. Danger, like sexual attraction, floods the brain with PEA. PEA also rises after ingestion of chocolate—which helps explain its status as a romantic gift.[17]

ROMANCE | ROMANCE IS DEFINED in *Webster's Dictionary* both as "a passionate love affair" and "something that lacks basis in fact." Living beings must perceive to maintain and reproduce themselves; sexually reproducing animals must also perceive to interact with each other. Given the universality of such perception, it should not surprise us that lying and loving often become intertwined. Organisms need each other to pass on their genes. They have evolved to play to each others' sensory recognition systems in order to do so.

Fertility signs—youthful skin, charm, social position—broadcast a given mate's aptitude to contribute genetically to the next generation. More than just fertility, these signs unconsciously convey reproductive potential, conforming, perhaps, to a future ideal offspring. In a famous set of experiments, babies judged composite photographs—not pictures of real individuals—as the most attractive. It is as if, in an evolutionary version of Plato's notion of an ideal realm of perfect forms, we are born with our own ideal of what constitutes beauty. Considering this dependence on signs, it is little wonder that men and women, having different, but interdependent genetic interests, have evolved duping tactics. After all, not just present life but future generations are at stake.

What we call love depends deeply on the natural trickery of living biochemistry. Odors, for example, that we find sexually attractive in the opposite sex, correlate with the potential for healthy immune systems in our offspring. In an experiment to discern any relationship between sexual attractiveness and pheromones, women sniffed T-shirts in which men had slept for several nights. The women were found to favor smells which correlated with histocompatibility proteins they lacked. In other words, we lust for those with complementary immune systems. So, too, there is speculation that romantic attachment—the special "light" or "aura" of the beloved in the eyes of the beholder—may be a pheromone-mediated imprinting phenomenon typical of mammals. Smell, connected to memory and removed from the language-processing centers of the brain, was probably the most important sense for our prehuman primate ancestors. Sexual activity implying proximity of the mate and exposure to his/her individual smells triggers intense, ancient mammalian attachment responses. From the early Cenozoic, when our mammal ancestors were nocturnal, females distinguished the males with whom they deigned to copulate by nasal appraisal and memory. Whatever else they may be, lust and love are ancient, chemically mediated biological phenomena.

Trickery among sexual organisms is rampant, but even organisms of different species, using the seductions of surrogate love, may pull the wool over each others' eyes. Male members of the *Silene* genus, a roadside flower, become infected by *Ustilago*, a fungus related to corn smut that "seduces" corn smut flowers into making seeds. These "seeds," however, although remarkably similar in appearance to the genuine plants' seeds, actually harbor propagules of the fungus. So, too, orchids produce natural sex pheromones and flower shapes that nearsighted male wasps and unsuspecting botanists mistake for female flower parts. A large percentage of brightly colored sexually breeding flowering plants (angiosperms) accomplish their own reproduction only by playing to the sex-based tastes of animals. This is not only interkingdom sensuality, but a highjacking of sexual reproduction: one species, finding that it pays to advertise, establishes a new, more effective thermodynamic pathway.

As seen above, seduction—sexual deception—even crosses species barriers. Means of cheating and means of detection, of promiscuity and fidelity, of maleness and femaleness and androgyny persist and evolve as long as they perpetuate life. Biologist Robert Trivers reports on the mathematical detection skills of some birds. They can count—at least enough to detect extra or too few eggs in their nest—having evolved such counting to protect their offspring and raise only their own young.[18] The ability to cheat or to detect being cheated, either by one's own or another species, selects for intelligence, discernment and critical thinking. As long as these means lead to the ends of new fertile offspring, sexual perception—and its associated intelligences and counter-intelligences—will continue to thrive. All is fair in evolution.

PLANET OF THE APES | WE ARE, BY FAR, the most populous primates on Earth. Yet we remain primates, mammals, blastula-forming animals subject to physical law and the path-dependent constraints laid down by our evolutionary history. [**PLATE 45**] All of our linguistic, cultural and technological abilities are rooted in the animal world; and animals, as we have seen, are intrinsically sexual beings, the meiotically fertilizing permutations, the embryo-forming descendants of cannibalistic survivors and their thwarted populations. It is useless to protest our separation from the rest of life when the microscopic image announces our likeness even to such watery sexual beings as amebas (See Chapter 4). Indeed, even our strong tendency to distinguish ourselves from other organisms probably reflects the requirement for sexually reproducing animals to discriminate themselves from similar others. We, like bonobos (pygmy chimpanzees), must recognize a potential mate amongst even slightly differing types of living forms. As self-perpetuating, sexually breeding beings, mammals must recognize and display themselves to other members of their own species. Since humans find self-confidence sexually attractive, and since self-distinction and gender display is a prerequisite to preservation, the discernment necessary to recognize one's species and the opposite gender probably contributes to our own view and each animal's view of itself as belonging to a unique and superior kind.

PLATE 45
**Ischial collosities (red Butt) of an available
young female mandrill.**
[Bruce Coleman]

Together, the social and sexual habits of modern apes hold clues to our sexual heritage. Sexuality may be broken into two great types, each corresponding to a given prehuman "morality" or species-based sphere of sociosexual propriety: the chimpanzees and humans on one side, the gorillas and orangutans on the other. These two camps differ greatly; the differences are reflected in the relative size of their genitalia and bodies. Male and female chimpanzee bodies, like those of humans, are nearly the same size. Also like human males chimps have relatively large genitalia and, indeed, they have larger testes and produce more sperm per ejaculation than men. One interpretation of the significance of large male genitalia is that humans were more promiscuous in the past, when our ancestors benefited from superior sperm delivery equipment. Currently, though, by human standards, chimps are much more promiscuous. Ovulating females, obvious by the swollen red skin around their vagina, copulate openly and often with many members of their group. All the males know when she is receptive because of the physiologic change known as estrus: her back end reddens and swells.

Neither orangutan nor gorilla females, by contrast, are promiscuous. They do not tend to expose their reproductive tracts to the attention of many potential male partners during periods of ovulation. Gorilla competition involves bodily aggression by the males. Sperm storage in scrotal-hung testes and delivery by thrusting and forceful penile ejaculation, which are of paramount importance to people and chimps, are less significant in the lives of gorillas and orangutans. That men and male chimps have longer penises, bigger testicles and produce more sperm than gorillas or orangutans is probably directly related to their tendency for more numerous sexual encounters.

Gorillas enjoy a "harem" breeding system in which one male, usually an older one with some gray hair—the silverback—dominates other males. This dominant "alpha" male has his pick of the females. He exhibits his

large frame, establishing dominance as he lords it over females and adolescent males. His erect, diminutive penis, only about an inch long, delivers relatively little sperm. Instead, male gorillas tend to be violent. Dominant alpha males prevent fertile females from mating with others. Conversely, in primate species where there is less difference in size between males and females, less male aggression, and thus less sexual possessiveness, competition is more apt to appear not at the level of bruising bodies, but of thrusting genitalia and swimming sperm. As long as male chimpanzees do not fight among themselves, but share females in heat, those shooting the greatest amount of fast-moving sperm closest to the egg will most often be favored. Large body size is an advantage among fighters, large genitalia among lovers.

Apes, including us, establish complex societies of couples and other gender relationships with rich interplay. Although evolution from common apish ancestors to us saw many changes in sexual lifestyle, our similarities to these closest relatives of ours are far greater than we choose to admit.

In laboratories, apes have been taught sign language and can recognize themselves in the mirror, a trait not displayed by monkeys. "When the apes"—and here Emory ethologist Frans B. M. de Waal is referring specifically to the bonobo (*Pan paniscus*) "pygmy" chimpanzees (blacker, with higher foreheads, smaller ears, and wider nostrils than their more familiar cousins)—"when the apes stand or walk upright, they look as if they stepped straight out of an artist's impression of early hominids."[20] [PLATE 46] We are distinguished by traits that include our upright posture, nearly hairless skin, large brains, persistent sexual readiness seen in our distributed period of estrus, relatively large male genitalia and the complex use of larynx, palate, throat, lips and tongue in speech. Comparing ourselves to our closest relatives provides us with evolutionary insight into our apish past.

Orangutans are loners of the Borneo forests. Sadly they are now primarily restricted to zoos. These expressive-faced apes are about two-thirds

PLATE 46
Erect-penis bonobo (cf. Plate 53).
[*Frans Lanting/Minden Pictures*]

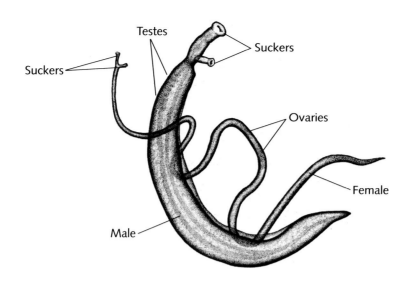

Testes

Suckers

Suckers

Ovaries

Female

Male

the size of a gorilla, with the male approximately twice as large as the female. Except when they mate, the pair enjoys little contact. In comparison, the less erect and much larger gorillas (*Gorilla gorilla*), whose habitat is equatorial Africa, live in small groups with one male and several females. Although more social than orangutans, sexually available female gorillas are usually available to only one—the huge, physically dominating adult male silverback or alpha who is apt to punish adolescent interlopers or any straying female who copulates with someone else. Finally, the most social of the great apes are the chimpanzees. These apes—one of whose two main forms, bonobos, displays the most overt sexual behavior—are in some ways the most human-seeming of the great apes. Bonobos mount each other normally during the course of the day under circumstances that might seem unsexual or socially inappropriate to other apes. Mounting and mating are inspired by such apparent irrelevancies as the receipt of an orange or the placement into their dwelling of a large, curiosity-arousing cardboard box. From the sexual behavior of bonobos and "normal" chimpanzees (*Pan troglodytes*) along with modern humans (*Homo sapiens*) we can deduce our likely sexual history.

PLATE 47
Schistosomes (Phylum Platyhelminthes). The flukes that wreak havoc with the lives of tropical people are permanently stuck in the mating position. As hermaphrodites, they infest the liver and produce hundreds of eggs, which they shed into the blood and other tissues of their victims.
[*Christie Lyons*]

DOMINANT FIGHTERS TO CHILDLIKE LOVERS | Sexual dimorphism, traits that mark differences between male and female organisms, is common in plants and most animals. Even in schistosome worms, liver flukes that spend most of their lives in the copulating position while shedding fertile eggs into the bodies of their victims (us), the "functional male" (the fat worm in **PLATE 47**) differs from the "functional female" (the skinny mate in **PLATE 47**). This sort of "functional" sexual dimorphism seems strange because each individual worm, both the fat sperm-producer and the skinny egg-producer, carry both kinds of sex organs (ovaries and testes). Although both are hermaphrodites, since each mate only uses one set of its sex organs at any given time, each in the mating act behaves as either a functional male or a functional female and look different as a result.

Sexual dimorphism is common in animals generally. In people sexually dimorphic traits include weight height and hair pattern differences [**PLATES 48, 49, 50, 51, and 52**] Among our ancestors, there have undoubtedly been the human equivalents of the dominant gorilla males with silver fur on their backs. Even now, our adult males relative to females display dimorphic dominance characteristics like those silver-haired gorillas. These traits include large size, solid musculature useful in jealous combat, male pattern baldness, coarse hair, dark hair and/or skin, low voice, gruff manner, beard, mustache, side burns and patterned streaks of gray hair. Even though our genitals imply a promiscuous chimp-like past, pre-human hominids such as *Australopithecus robustus* may have been hunted to extinction by less dimorphic, but more aggressive, smarter early humans, such as members of the more recent species, *Homo erectus*, who dwelled from 1.3-0.5 million years ago.

Yet because we are not typically harem breeders such dominance characteristics, historically associated with the most physically and socially powerful hominids give us mixed messages of attractive power and unattractive tyranny. Our transition from apedom is far from complete. In us, for example, hair retains its ancient status as an ape hieroglyphic. Greasy

---

Following pages:

PLATES 48
Angry man.
[International Stock Photos]

PLATE 49
Angry gorilla.
[Frans Lanting/Minden Pictures]

PLATE 50
Emperor tamarin (bearded); *Saguinus imperator subgriscens.*
[Tim Laman/Wildlife Collection]

PLATE 51
Soldier in bright red guardsman's uniform, sporting handlebar moustache.
[Comstock]

hair, coarse hair, dark coloration, body hair, large stature, gray hair and male pattern baldness are all apish body signs of maleness, aggression and dominance. By contrast soft skin, smooth skin, light and downy hair are signs of youth and childlike femininity—less threatening and more disarming. "Cute" organisms are more likely to attract attention and care.

Humans and other mammals respond to youthful traits such as large eyes, proportionally big head, soft, cuddly skin and small teeth. These traits are just those our offspring who require parental care for survival possess. Cuteness is especially important. Humans retain, even into adulthood, traits which in ancestral primates did not linger beyond childhood. We resemble young apes more, in other words, than we do adult apes. Our ancestors apparently underwent a change in embryological timing, being born earlier at a less mature stage. Today our large brains continue to grow and expand outside instead of inside the womb, where their large size has historically created pains and problems during childbirth. A suite of characteristics (large heads and eyes, small canines, lifelong curiosity and relative hairlessness) are neatly explained by the thesis of neoteny: the tendency toward childness. Human adults, who become more apish with maturity in the sense that they develop more hair, are nonetheless far less hairy than ancestral adults. Even modern leaders such as President Bill Clinton tend to be more baby-faced and smoother-shaven with a pleasanter manner and softer hair than their predecessors. As neotenous leaders their features well illustrate the trend to become less threatening and "cuter."

Societies act as units. The power politics of large, physically stronger males and smaller, weaker women and children evolved as human populations increased. Today, men with the look of the dominant hominid in the past—large frame, low voice, coarse hair, baldness, facial or gray hair—are as likely to command fear as respect. As our hominid ancestors grew more populous, they began more frequently to encounter strangers. In cities full of strangers, the look of dominance is also one of potential criminality and violence. Ameliorating threats of danger is one reason that men have taken

PLATE 52
**Sexual dimorphism. Jumping spiders,
Victoria mating ritual.**
[Mark Moffett/Minden Pictures]

to shaving their faces, a custom begun by the Macedonian military. Combing and grooming the hair to make it look softer, hiding baldness, darkening gray hair and speaking in soft, civil tones all mitigate against the ancient rages of the dominant male. The custom of private urination is probably also a response to populous hominid living. So probably too is the masking of body smells stronger at close range than they were in our more sparsely populated ancestral environments. Contemporary cultural responses have insisted that aggressive males temper their dominant behavior as they leave more offspring, in marked contrast with the dominant patriarch who once universally commanded respect and sexual attention. The movement away from alpha male brutes in the human species was thus doubly motivated: by the importance of moderating aggressiveness brought on by increased population density and by neoteny.

DEADBEAT DADS | EVEN IN THE ABSENCE of material culture, clothes and language, typical female and male behavior can be notoriously mutable. Throughout all reptile and most mammalian social organization, no lasting bond exists between father and child. Males compete to inseminate females and leave them to care for the young. The female may choose males more devoted to her and her young, but male connection to offspring is most often absent or indirect. A common theme is the provision of food. For example, sperm-containing edible gifts called spermatophores represent up to forty percent of the body weight of some male insects. By enticing females to eat the nutritious portion and store the sperm for later use in a special organ (the spermatheca), males exchange food for genetic representation in future generations. In a colorful extension of the food-for-sex theme, the black widow spider and the praying mantis both devour not the spermatophore but the mate himself after he has performed his sperm-passing function. Even macho bats, flying mammals far more closely related to humans than are spiders and mantises, produce spermatophores.

Nonetheless, an inevitable tension arises as males retain the physical capacity to impregnate many females—to "stray" from infertile nursing or lactating women to fertile not-yet-pregnant women. Given the great vulnerability of human infants, this dramatic tension seems to have accompanied humanity since our inception. Indeed, the potential for fathers to stray may be reflected not only in the worried female mind but in her body itself. Women lack the vulval and anogenital body markings typical of female primates in heat. Estrus, the swelling and pinkening of the female vulva around the time of ovulation, is a sex lure found in primates including chimpanzees and baboons. [PLATE 53] Estrus signifies that the female is ovulating. Such external signs of sexual readiness aid conservation of energy by members of a species who accurately time the act of mating. But for females requiring commitment, advertising monthly fertility tempts males to jump ship. Tribes in which females had evolved to lose or disguise obvious signs of fertility were less likely to contain sexually distracted absentee fathers. Synchronized menstruation, which often occurs to women living in close proximity to each other, as in college dormitories, would also have discouraged males from leaving one female to find a more fertile female: all the females in a given protohuman primate troop would be fertile on the same days, effectively making it impossible for males to carouse.

Women today show no obvious signs of ovulation; estrus is cryptic. Moreover, enlarged mammaries—as opposed to other primates with nipples, but breasts that swell prominently only during nursing—probably confused early men on the lookout for fertility. Men tend to show less erotic interest in females who are unlikely to conceive—older females, unhealthy females, and pregnant and lactating females. The behavior of lions, tigers and bears we judge to be far worse. The murder by aggressive males of nursing cubs converts their non-ovulating mothers into available sexual creatures newly capable of insemination by the excited murderer.

Prominent breasts and concealed estrus, which made it more difficult

PLATE 53

Estrus. Stretching female bonobo with prominent vulva.

[*Frans Lanting/Minden Pictures.*]

for men to detect ovulation from external signs, probably made them more likely to devote themselves—if at times possessively and controllingly—to a single female at a time.[21] Cryptic estrus, in other words—a kind of victory for women in the "battle of the sexes"—helped induce our male ancestors to form bonds with their children. It increased the rate of father care. While men do not typically raise offspring alone—as do, for example, males of the lily pad-dwelling African jocana or Jesus birds—one can easily see that fathers played less of a role in child rearing for our ancestors and that, far from being a recent phenomenon, sex roles have long been in flux.

LAUGHING HYENAS | INCREASED CHILD REARING by human males is just one of thousands of examples of the potential for evolutionary change inherent in the two mammalian genders. In a recent lecture, the articulate feminist writer Susie Bright mentioned a female friend of hers to whom she was attracted and who was in the process of receiving hormone treatments as part of a medical makeover to become a man. The friend remarked that the hormones were driving her crazy. "I'm so horny I feel like raping Bo Peep." This comic comment alerts us to the important role that hormones play, above and beyond our genetic constitution as male or female, in making us masculine or feminine. The so-called war of the sexes is more like a skirmish in a long campaign since the combatants and battlefields change so quickly. The distinct physical differences and psychological tendencies between men and women are mainly the result of hormones working on a unisex-looking embryo. Hormone levels, past history, social behaviors and many other factors influence what we so often think of as fixed: the gendered body and its sexual behaviors.

Some species of mammals show striking gender reversals far beyond our human experience. One example suffices here: the "masculinization" of the laughing hyena. Of the four extant species of hyenas only the giant-clitorised spotted hyena is amenable to scientific study, since it alone is not on the endangered species list. The spotted or laughing hyena (*Crocuta crocuta*) is an aggressive mammal with impressively "masculine" females (**PLATE 54**). All-female packs of these razor-teethed creatures reduce a zebra to four hooves in twenty minutes. Their feces are white since they chew and digest bone.

Until the 1990s, no one had successfully determined the gender of these creatures for the reason that all of them, both males and females, apparently sport pendulous penises. Scientists now know that the "female penis" is in fact her enlarged clitoris, whereas the male organ is smaller and wider. The females, who are impregnated by smaller, squatting males with very long penises, entirely lack vaginas. Unlike other placental mammals, the babies, usually twins, must make a U-turn through the birth canal to come out through the clitoris. Painfully, birth transpires along the length of the urethra, which runs inside the giant clitoris. The meatus, or opening at the end of the urethra, rips to make way for a four-pound hyena infant. Many mothers die during childbirth; many first cubs are stillborn. But the species survives. Why? Apparently because teams of aggressive female hyenas hunt with great efficiency and breed successfully despite their high death rate. Yet in no way are laughing hyenas unnatural.

Nature tends toward variety in accord with the Second Law; yet insofar as it can maintain its form, reproducing itself, sexually or otherwise, it breaks down gradients, also in accord with the Second Law. Sexual breeders must come together socially, which naturally leads them to evolve means of reducing gradients in groups. A first time hyena mother dying in childbirth will not pass on her genes yet, in the aggregate, her vicious hunting sisters will outcompete less aggressive predators. Despite its high mortality rate, the strategy of hyena efficiency has led and will lead to more hyenas.

A tiny chemical change can have powerful evolutionary effects. The mother's placenta of these masculinized hyenas synthesizes testosterone

in utero. Like transgendered humans, vaginaless female hyenas may seem strange. But life shows far more variety in the sexual realm, both within and among species, than our straight-and-narrow view of normalcy might suggest. For their own species, hyenas are perfectly normal.[22] Just as cultural fashion changes so that women now wear pants and sometimes even neckties, while men may wear long hair or earrings, genders evolve and may even "reverse" over evolutionary time.

PLATE 54
Female hyena with large, penislike clitoris.
[Frans Lanting/Minden Pictures]

six

COME TOGETHER:
THE FUTURE OF SEX

We only have two sexes; we can't conceive of a third:
all we can imagine are interesting
combinations of the two.

—*Robert Garrels*

SUPERORDINATION AND THE METAMORPHOSES OF CROWDS | CROWDS are strange. Whether at a rock concert, sporting event, or battle, crowds seem to have a mind of their own. But these human gatherings are nothing compared to the behavior of certain crowds over evolutionary time. Throughout evolution, formerly individually reproducing organisms have congregated into larger collectives more apt to function and reproduce as collectives. Over time, the many can become the one. New kinds of life evolve from association. In some cases, members of the collectives stop reproducing as individuals or as pairs. In some cases, each may act as part of a new larger organism and, as such, the collective becomes more effective at reducing gradients than are the unassociated members. In colonies of microbes, nests of social insects, dens of naked mole rats and cities of humans, the incessant need for sexual gratification is sacrificed or even lost as thermodynamic efficacy of the greater whole is gained. The collective competes with other dissipative genetic entities and leaves more offspring in the eternally changing evolutionary wilds than do the individuals independently.

Emotions are intrinsically social. But so are animal bodies. Intimate and regular contact with mother, sibs, colonies, mates and other members of one's species is essential for maturation and eventual reproduction especially in primate mammals. Blushing, triggered by exposure as the unintended center of attention, is marked by a reddening of the face that reflects increased capillary action and blood flow to the head. Sparked by embarrassment in social situations here is one striking example of individual physiology intimately connected to group physiology.

Desire, in psychoanalysis, refers to the effort to find a missing object (or to rid oneself of an extra object) that will make us feel complete. But because we are open systems in an ever-changing, dissipative universe, whose nature is linked to our perception of a forward direction in time, no possible static possession or removal of an object, sexual or otherwise, will truly make us whole. It is the movement to fulfill our desire—some call it lust—rather than arrival at any destination of final satisfaction that is consonant with the Second Law of Thermodynamics.

Recombination is a fundamental process of life. Words are combined and recombined to form language. Images are combined to form art. Organisms with genes are combined in ever-changing arrangements to form unique offspring. [PLATE 55] In sex, imitation is the highest form of flattery, and plagiarism is a form of fidelity. Through symbiosis and sex, organisms appropriate the hereditary skills of others. Appropriation of such skills sometimes saves lives, while of others sex merely promotes quirks. But because of the hereditary, mnemonic quality of life, newly recombined organisms—like all others—copy, with varying degrees of success, their own form.

The dialectic between conformity and novelty, between repeating order and promoting diversity, is central to life.[1] Whereas socially accepted standards of sex in the missionary position belong on the side of conformity, technological experimentation by genetic engineering belong on the side of novelty. We see that variety and mixing are a basic tendency of the universe in accord with the Second Law of Thermodynamics and the forward (-seeming) direction of time. If verified, cloning sheep would be an example of enhancing life's confirmatory potential via the human novelty of ever-evolving technology. But the subtleties of the environment in which embryos grow and develop outside the mother's body are such that even "perfect" cloning will not ensure identity. As the philosophers have pointed out, a perfect copy is the original. In real life, the variety-generating tendencies described by the Second Law are impossible to eliminate. Identical twin children raised together—as close as we can come to perfect copies of humans—show major differences in the neuronal architecture of the brains due to their different experiences.[2] The natural variety-engendering and fidelity-maintaining processes of life, coupled as sex and reproduction in

Symbionts
autopoietic entities
*autopoietic entities of different
species mutually attractive*

Bionts
eukaryotic organisms
*complementary genders
mutually attractive*

Biont 1   **Biont 2**

Recognition

Merging

DISSOCIOPHASE   DISSOCIOPHASE

Integration

HOLOBIONT
lichen

Association

ASSOCIOPHASE

Dissociation

Fungus   Algae

(bionts)

Gamont 1   **Gamont 2**
Gamete   Gamete

Recognition

Mating
Fertilization

HAPLOPHASE   HAPLOPHASE

Karyogamy
(nuclear fusion)

Zygote (fertile egg)

ORGANISM
diploid

DIPLOPHASE

Meiosis

Sperm   Unfertilized Egg
(spores, cysts)

animals, interact in many ways to produce the splendid complexity of evolutionary history. In today's world this extends to the incorporation of computers into our sex and social lives.

CONVERGENCE | A PERSISTENT OBSTACLE to a truly evolutionary picture of our universe itself, ironically, has an evolutionary explanation. The obstacle is our culture's widespread assumption that human life is the purpose of creation and that all other organisms, soulless automatons that they are, were created for us. This self-aggrandizing view is strongly reinforced by the traditional idea that we were made in the image of God. We are not special. Earth is not the center of the universe. The matter of our bodies is not unique. We are not composed of pixie dust. The same hydrogen, carbon, phosphorus, oxygen and other atoms found in outer space and in all other life comprise us. We play a smaller role than do trees in the thermodynamic gradient-reducing systems of life cycling material into its own form on Earth. We are no more the "highest life form" than we are the "chosen species" for which all others were created. Nor has evolution or nature somehow "ended" with us.

Indeed, more than ever, the evolutionary consequences that have affected other crowded, rapidly reproducing populations of organisms—and effected their organization into "individuals"—now appear to be occurring to us. Convergence is the biological term for the separate evolution of similar features. Convergence occurs when separate lineages, under similar environmental stresses, independently evolve the same sorts of body parts or behavior. Although whales, extinct marine reptiles, and tunas, for example, all have hydrodynamically shaped bodies, subcutaneous fat reserves and sleek skin, they share no immediate common ancestors. Rather, the ancestors of these swimming beings were air-breathing mammals, land-dwelling reptiles and sea-swimming fish, respectively. Wings evolved similarly, not because insects, bats and birds share immediate common

PLATE 55
Cyclical symbiosis (left) compared with sexuality (right). Both involve singleness (haplophase, dissociophase), recognition, merging (fertilization, symbiogenesis), integration (associophase) and dissociation (meiosis).
[Kathryn Delisle/José Conde]

ancestors, but because all were forced to fly to feed and survive in relatively uncrowded aerial environments. Hydrodynamic shapes for thriving in the watery milieu and wings both provide examples of evolutionary convergence.[3]

As in certain species that preceded us, including some protoctists, slime molds, insects (termites, bees, ants and wasps) and many others, the number of our individual reproducers, as human groups more successfully organize, is on the wane. We foresee, in fact, humans evolving into a species composed of populations in which most members do not reproduce.

While rarer than standard speciation into divergent species, superordination—social aggregation to form newer, larger organisms—is not freakish or unique. Precedents abound. When merging members came from different species, we speak of hypersex—as when rapidly dividing bacteria merged to become more slowly reproducing cells with nuclei. When the merging members are conspecifics—that is, when they belong to populations of the same species as do, for example, members of slime molds such as *Dictyostelium*—we speak of superordination. Superordination in the group physiology of what we think of as societies of discrete individuals can be surprisingly complex. Social ants, bees and termites have developed into well-housed superorganisms which, inside their hives and mounds, maintain humidities over 90% and ambient temperatures of 18-24°C. Other individuals cooperate to display unique behaviors unavailable to individuals. Japanese honeybees detect the pheromone of hornets that prey on their nest. As the predatory wasp scout approaches, as many as 500 honeybees congregate outside the entrance to the nest to form a honey-bee ball around the wasp. The bees tolerate temperatures up to 122° F. Vibrating their wings, they rapidly raise their aggregate temperature above 115° F, enough to cook the wasp! The ambushed scout perishes when the temperature reaches 114° F.

The common myth that we are superior to all other organisms because we have intelligence and culture, whereas all other organisms are blind automatons wholly genetically determined, is easy to refute. By convergence, populations superordinate. As our global population increases, we reorganize, birth rates reduce and we become more technologically interconnected. Our society has begun to decouple sex from reproduction. Religious strictures, contraceptives from condoms to natural remedies and synthetic "day after" pills, abortions, rising infertility, sperm banks and the science of cloning already have snipped the biological cord tying sex to reproduction. Masturbation inspired by glossy magazines and X-rated videos or phone sex engender sexual pleasure in the virtual absence of the possibility of pregnancy. Although limited now to a small percentage of Earth's population, cyberspace is a further extension of an evolutionary trend toward the decoupling of sex and reproduction. We suspect this trend of separation will continue as more and more people jostle for space, food and energy.

Cells evolved into bodies. Insects evolved social units as hives and termite mounds. The "rewiring" of sexuality relates to self-regulation of human populations. Biologists call formerly sexual populations apomictic. People, falling in love with machines, may be heading toward apomixis. Just as the inherited tendency of all the cells of our body to grow is arrested under the discipline necessary to make healthy, nontumorous tissues, so the inherited tendency of humans to indulge in sex for reproduction is being mitigated. Our populations are evolving to become less like dangerous growths and more like well-behaved neural tissue of the biosphere. The sexual attraction which induces us to seek mates at clubs, dances or via newspaper classifieds or on-line forums easily modifies to become strictly social lubricants with few reproductive consequences.

Each of us is not necessarily aware of the social function of our sexual impulse as it works to incorporate us into the hyperbrain. As Samuel

Butler wrote of the component cells of our bodies, they "unite to form our single individuality, of which it is not likely that they have a conception, and with which they have probably only the same partial and imperfect sympathy as we, the body corporate, have with them."[4]

ROYAL RATS | IF IT HAPPENS, we will not be the first mammalian species to shut down sexual reproduction in most of its adult population. This crowd behavior has already occurred in *Spalax*, African burrowing rodents called naked mole rats. Among the ugliest organisms on Earth, these wrinkly pink "saber tooth sausages" live and sleep in colonies underground whose entrances are guarded by sentries. [**PLATE 56**] Only the huge queen and two or three breeding males reproduce. Other males and females occasionally indulge in "anogenital nuzzling" or even climax in sexual intercourse, but five of seven males, if they ejaculate, produce only dead, infertile sperm.[5]

The infertility of most naked mole rats appears to be hormonally mediated. Naked mole rat queens—not their male consorts—have the highest blood levels of testosterone of all colony members. Like many other mammals, including humans on corporate ladders, naked mole rats form "dominance hierarchies," or "pecking orders." The development of "dominance hierarchies" in which particular individuals assume distinct roles is intriguingly reminiscent of differentiating cells. Populations, whether of animals or cells, gradually specialize for different functions. Specialized populations, like the cells that differentiate into animal bodies, use energy more efficiently, making them more effective than less organized populations operating in the same area.[6]

The male steroid hormone testosterone acts as a differentiator. [**PLATE 57**] When the queen is removed from a naked mole rat colony, testosterone blood levels skyrocket in both breeder and nonbreeder mole rats. In one test by Chris G. Faulkes of the Zoological Society of London, researchers

PLATE 56
Naked mole rats, *Spalax*, root-eating communal mammals.
[*Gregory G. Dimijian/Photo Researchers*]

tagged 100 males and studied their crawling behavior as they made their way through narrow Plexiglas tunnels. Those males that passed on the top were given the highest scores. When the queen was removed, the dominance ranks were overturned. The colony can not tolerate queenlessness, and soon an ordinary sterile female began to develop testosterone-mediated queenly tendencies. Only one male, a large potential breeder, had scored higher than the queen-to-be, by passing on top even more often than she did when confronted with others in the narrow passageway. She murdered him.

The queen's behavior, as well as her testosterone levels, apparently inhibit the sexual development of other members of this "reproductive dictatorship." Similar testosterone-lowering effects have been observed in primate societies among fighting males. After beating a dominant male, victors sometimes show elevated levels of testosterone, while losers may experience significant declines. Somehow, social wrangling and the new status in the dominance hierarchy translates back into physiology. Human drug use and bio-active chemical compounds, often considered to be a meaningless exercise in apathy or immorality, can be interpreted in the light of their social effects. Connecting, altering and often debilitating individual humans, drug use alters human populations. Drug addicts probably sire fewer offspring than abstainers. Marijuana, for example, whatever its recreational allure and possible health handicapping or benefits, temporarily lowers testosterone levels. The potential of drugs to reduce population size is obvious. Indeed, the destructive character of widespread human drug use is apoptosis-like, predictably leading to human losses but at the social rather than the cell or tissue level.

As close-living communities merge into larger units, evolution occurs by branching and fusing. Hermit mole rats, solitary termites, isolated ants and abandoned children do not tend to live long. Reproductive and nonreproductive organisms coexist in efficient social groups that function as

---

PLATE 57
Crystallized hormones. These molecules in solution circulating in our blood stream have profound effects on life and sexuality.
[Lennart Nilsson]

units. Hormones that communicate between cells in tissues of a single body are co-opted to organize hundreds of termite or mole rat bodies into collectives. Male naked mole rats who inseminate the queen are the social equivalent of testes. Just as embryonic stem cells from a human fetus have totipotency—they can still reproduce without limit and have not been exposed to factors both specializing them and curtailing their reproduction—so rodent ancestors to the naked mole rats probably could all reproduce. Our own bodies, composed of trillions of nucleated cells differentiated to form tissues, lose their reproductive capacity very early in embryonic development. It is as if embryonic animal development replays, in utero, the high drama of moving from a society of individuals each of whom can reproduce to a collective in which only a few select members leave offspring.

SPERM LOSS AND DENSITY DEPENDENCE | SIGMUND FREUD stressed an irony of modern life, namely, sexual repression. But sexual repression, that great source of personal consternation, is foundational to culture. Limited energy, redirected from the business of mate seeking and love making, is "spent" in socially productive ways.

> The sexual life of civilized man is seriously disabled....it sometimes makes an impression of being a function in the process of becoming atrophied, just as organs like our teeth and our hair seem to be....the importance of sexuality as a source of pleasurable sensations, i.e. as a means of fulfilling the purpose of life, has perceptibly decreased. Sometimes one imagines one perceives that it is not only the oppression of culture, but something in the nature of the function itself, that denies us full satisfaction and urges us in other directions. This may be an error; it is hard to decide.[7]

Several lines of evidence suggest that sexual activity is becoming less important for reproduction, even as it becomes more important for communication. Sexual activity, welding connections among increasingly interdependent human groups, no longer necessarily generates so many offspring. Sperm count around the world appears to be declining. The dramatic loss in average number of sperm per ejaculation has been attributed to environmental toxins which mimic the effects of sex hormones. A chief culprit may be the manufacturing byproduct dioxin with its estrogen-like (female hormone-like) effects. Concentrated by cows, and spread in part through milk, estrogen-like compounds may be responsible for an observed planetary decline in virility.[8] Gender disorder in fish, for example, the feminization of male carp, has also been ascribed to industrial chemicals that decrease reproductive rates.[9]

From the beginning, nature has incorporated accidents. We are tempted to believe in a standard, "normal" two-gender sexuality, but over the long haul, genders change. Imperfections, developmental abnormalities, kinks, sports, "monstrous" new forms of behavior, including reproductive behavior, arise and persist as sexual organisms undergo social transformation. A human named Emma was born with a penis-size clitoris and a vagina, and as a teenager she engaged in "normal" heterosexual activity with a number of girls. She continued sexual activity with women, even after marrying a man at the age of nineteen. When informed by urologist Hugh H. Young at John Hopkins Medical School that her wish to be a man could be medically realized, Emma replied:

> Would you have to remove that vagina? I don't know about that because that's my meal ticket. If you did that, I would have to quit my husband and go to work, so I think I'll keep it and stay as I am. My husband supports me well, and even though I don't have any sexual pleasure with him, I do have lots with my girlfriends.[10]

Although Emma was exceptional, she was certainly natural. Necessarily social, sexual attractions, repulsions and repressions exist along a continuum. Over time new norms, new standards of mating and fertility, new behaviors and chemical cues evolve. Just as our close ape relatives, the bonobos' tree-swinging sex in which children join may seem perverted to us, so our automobile advertisements and movies may seem oversexed in retrospect when viewed by our primate descendants. The most successfully reproducing organisms create the biggest messes which, in turn, create the greatest problems for the continued growth of those same organisms. Our success at sexual reproduction has set up the very conditions for the diminution of that sexual reproduction.

HORMONES AND PHEROMONES | WAFTED INTO THE WATER, soil or air by one organism, allelochemicals have a meaning to others of the same species. Allelochemicals can be thought of as "ecological hormones." Pheromones, examples of allelochemicals, are social chemicals that signal between members of the same species. They include molecules produced by bodies of one gender that elicit physiologic and behaviorally measurable changes in members of the complementary gender. Hormones released in an individual signal the body of that same individual. Hormones, such as testosterone and estrogen, which are found in both male and female mammals, probably originally functioned as pheromones long before mammals evolved. They signaled to organize societies of cells. Pheromones that at first regulated attraction and behavior in protoctists probably became hormones that regulate growth within the bodies of single individual animals. As cells stayed together after mitotic reproduction to become organisms, chemicals regulating behavior among members of a population of those cells become the chemicals that regulate behavior among cells of a body. The process can also work in reverse: as organisms merge to become societies, the hormones inside them can "bleed" back out into the social environment.

Testosterone, for example, the same steroid that makes boys develop beards, low voices and body hair as they physically mature into men, also serves as an attractant—a pheromone—between mature animals. For example, a male lamprey, an aquatic fish-like vertebrate with teeth but no jaws, can attract a female lamprey by emitting testosterone in the minute amount of 29 picograms per milliliter. Young male mice (Mus musculus) exposed to the testosterone-containing urine of grown males mature into lighter-weight adults with smaller reproductive organs. Female mice, however, are not affected. Young female meadow voles (mice-like rodents, Peromyscus leucopus noveboracensis) develop undersized ovaries if they are, at a critical period, exposed to adult male urine. By contrast, researchers found that female urine had little effect.[11]

Such experimental evidence suggests that no universal language prevails for hormones and chemoreception: the same chemicals evolve different effects in different animals under different conditions. We blame worldwide sperm count diminution on the peculiarities of our own technology and pollution and on environmental degradation. But pollution is absolutely natural and, indeed, an inevitable consequence of any rapidly growing life form. Because all organisms have metabolic input needs that differ from their gaseous, liquid, and solid outputs, all will die of their own wastes if their populations grow too rapidly in a limited area without metabolically diverse helpmates. Organisms, composed of formerly independent selves, appear at ever-higher levels on the ecological stage. They do so in part because the higher-level entities have resolved problems of self-poisoning accruing from overcrowding.

Diminution of human sperm production is a natural example of reproductive "braking" due to self-poisoning caused by runaway reproduction. A given type of organism can grow only so rapidly before being confronted with its own accumulating waste. Dioxin is a contingent, technological product that could probably not have been predicted. But the general appearance of reproduction-lowering influences on the human population is inevitable because of our rapid growth. Dioxin and many other pollutants interfere with our reproduction. The pesticides we use to control crop damage also damage us when we eat those crops. The antibiotics we use to kill disease bacteria select for more virulent forms of those same disease bacteria.

Ecology, from local ecosystems to the planetary biosphere, depends upon, and naturally foments, diverse pathways to accommodate sunlight hitting Earth's surface. Populations of certain organisms spread quickly for awhile as they store and release energy, but those best able to persist function as wholes, as complex organizations with many metabolic interconnections. It is not difficult to see that the current motley free-for-all of human population growth could be replaced by more orderly forms of expansion.

Sexual intercourse now prerequisite to reproduction may be co-opted for other roles in future human collectives. Medical techniques today fertilize ova without orgasm or erotic pleasure. Even now we know that orgasm need not be experienced only in the genitalia. Nongenital orgasms have been documented and analyzed in women who suffer damaged spinal cords. These women may feel orgasmic pleasure on the shoulder, chest or chin. Barry R. Komisaruk and Beverly Whipple of Rutgers University have shown that the vagus nerve, present in early vertebrates, is a "primitive and more commonly found pathway for orgasm."[12] Pleasure promises performance of the actions that lead to reproduction. But, as population size plateaus, the sex leading to reproduction, now excessive, becomes freed or let loose of other tasks that, from the standpoint of reproduction, may appear perverse. As one man's meat may be another's poison, so the normalcy for one species may have begun as perversity, or (to speak less moralistically) as sexual creativity, in its ancestors. Our human ancestors, as attested to by the evidence of sperm competition, may well have been more bonobo-like—more promiscuous and sexually active—than today's humans. Evolution rarely stands still. As we lose our personal reproductive prerogatives to group integration, individual sexual pleasures are expected to be redirected toward the maintenance and expansion of new social organizations. That individual sexual pleasures lead to unpredictable excesses on which evolution can potentially act, separating the ancient link between sex and reproduction, is nowhere more glaringly revealed than in the figure of the Marquis de Sade.

THE SCREAMING BLUE MARQUIS: SEX IS EVERYWHERE | THE OCEAN kisses the sand. The sky touches the horizon. Rappers sample, inserting the licks and riffs of other artists within the ecstatic identity of their own works. Some cave boy must have realized, long before Freud brought it to the attention of his medical colleagues, that almost anything can have a sexual connotation. When, in our materialistic, increasingly global and capitalistic society, mention of sex is made explicit, product sales increase. The powerful dynamic works because of an instinctive, biochemically modulated taste for sex which, to put things imaginatively, in its bondage to reproduction works for a pimp called the Second Law.

There is probably no more prolix pervert in the history of sexuality than the Marquis de Sade (1740-1814), who claimed that sodomy was not only the most pleasurable means of birth control, but so blissful that should the mass of humanity discover it, *Homo sapiens* would rapidly become extinct—and this would be a fitting end. The Marquis' books were so foul that his own son burned them. Yet in his defense commentators have pointed out the Marquis was peculiarly consistent and was only trying to beat nature at her own game. They stated that if any fingers were to be pointed they should be done so at the hypocritical society in which he lived and not at the obsessive Marquis. The feminist Simone de Beauvoir (1908-1986), author of *The Second Sex* and paramour of existentialist philosopher Jean Paul Sartre, even held that the Marquis was a "great moralist."

The Marquis was obsessive. He carried with him a pillbox full of candied Spanish fly, the "aphrodisiac" urogenital irritant used to induce copulation in animals. Hiring prostitutes even in the first weeks of his marriage, the Marquis was eventually jailed for abducting, whipping, and pouring molten candle drippings into the whip-induced wounds of a young woman named Rose Keller. At his bidding, his manservant Latour engaged in sodomy with him. The Marquis' wife, prior to becoming a nun, participated in the orgies organized by her depraved husband. He eloped to Italy with her sister, the canoness of Launay. The Japanese author, Mishima, in his play about the Marquessa de Sade, even imagines an orgy involving the Marquis' spouse whipping a nude boy hanging from a chandelier.

Jailed more than once for such activities, perhaps the only thing the Marquis liked better than directing his sexual parties was writing about them. All his books were banned in February, 1784. Five years later, in 1789, during the French Revolution, he was transferred from the Dungeon of Vincennes to the Bastille in Paris. The "evil Marquis" remains history's most literate sexual villain, lending his name to the term "sadism"—even though, according to some of his apologists, his aesthetic consisted in provoking not just pain, but heightened sensations of all kinds.

One might think that the Marquis' behavior belongs on the outer rim of nature, an aberration so extreme that it could be produced only by the dementia of a warped human consciousness. Certainly the Marquis was aberrant, a sexual pervert and a criminal. Representing an extreme version of the admixture of violence and sexuality that may or may not be natural, he surely deviated from the human norm. But compared to nonhumans, the mischievous Marquis seems downright tame. Next to the actions, for example, members of certain spider species routinely impose on each other, the antics of the Marquis were relatively harmless. Lamenting the monotonous sameness of human sexual behaviors and practices, Steven Shaviro of the University of Washington points out that,

> Even our wildest S&M fantasies are all too often trite and formulaic.... Far from simply exalting cultural difference, we might do better to ask why human cultures aren't *more* diverse than in fact they are. From a biological standpoint, our sex lives are

exceedingly dreary. Other organisms are far more inventive. Consider, for instance, the bedbug (Climex lectularius). The males of this species fuck by stabbing and puncturing their conspecifics' abdomens. Every copulation is a wound. The victims of these aggressions, males and females alike, are permanently scarred; and they carry their rapists' sperm in their circulatory systems for the rest of their lives. As Howard Ensign Evans puts it: "the image of a covey of bedbugs disporting themselves in this manner while waiting for a blood meal—copulating with either sex and at the same time nourishing one another with their semen—makes Sodom seem as pure as the Vatican." Even Sade never imagined such a scenario. No Californian cult of the self here! We humans should be thankful that bedbugs regard us not as sex partners but only as food....There's no clear dividing line between body and thought, or nature and culture, just as there is none between the water and the land. Language and sexuality are not the clean, abstract structures the so-called 'human sciences' have long imagined them to be. Rather, they are forces in continual agitation in the depths of our bodies.... Our bodies join and separate: this is the mark of the *social*, whether in frogs or human beings or prokaryotes. To speak of human culture is much the same thing as to speak of a 'culture' of bacteria. Only those dazzled by Gutenberg's movable type, or by the concurrent figure of "Man," could ever have imagined otherwise. But now "Man" is on the verge of disappearing: he is gradually being erased, as Foucault puts it, "like a face drawn in sand at the edge of the sea."[13]

The world is a funny place. As we have seen, sexual reproduction is a byproduct of thermodynamic dissipation. Bacteria in their ancient furor of genetic exchange desperately and non-reproductively engage in sex for survival. A long and painful evolutionary history, two billion of years in the making, linked the imperative of sex to reproduction. There is nothing predestined or intrinsically "natural" about this linkage. It is a byproduct of the contingent history of protoctists, seasonally stressed beings whose survival depended on locked cycles of fusion and meiosis. These tiny sexual reproducers are our ancestors. Sex appears to be pleasurable to us because, linked to our reproduction, it is the way that we stave off thermodynamic equilibrium, thereby obeying, if not accelerating, the natural tendency of the universe to break down gradients as we move forward in linear time.

From this perspective, sex is secondary to thermodynamic dissipation. In the short run—in our individual lifetimes—we produce entropy by maintaining our identity, which necessarily entails the elimination of liquids, gases and solids through our orifices. In the long run we ensure the production of entropy by mating, which produces new organisms like us that continue the special form of dissipation known as life in the next generation. But the Second Law of Thermodynamics, rather than sex itself, is the physical basis of our evolutionary focus on copulation and other sexual acts. The linkage between sex and reproduction, in other words, is contingent. If other dissipative processes are available to replace it, it can come undone. But the unlinking of sex and reproduction is never a simple reversal. Life always retains clues to its tortuous and weird history.

CYBERSEX | TECHNOLOGY, which has been drawn into the sexual orbit on many levels, has of course become increasingly important for the survival of global humanity. As technology and standards of living improve, population growth begins to decline. Sociologists call this "the demographic transition." The demographic transition probably provides a human example of the tendency of mammals to devote more resources to fewer offspring.[14] A mammal infant suckling at the breast enjoys a greater chance of survival than a baby amphibian, reptile or fish. There is an evolutionary tension between the "shotgun" approach of producing many offspring and the less wasteful, more calibrated approach of producing fewer, better cared for offspring. Not only do we seem to be moving from the former to the latter but the rise of technology exacerbates this unlinking of sex and reproduction to the vanishing point: many of us now produce no offspring but instead refocus our sexual energies.[15]

Cyberspace, as it exists now on the on-line networks, is only a harbinger of the screenless, image-based, computer-generated worlds of future long-distance communication. More than downloadable photographs and instantaneous chat, future electronic worlds will be entered, reconnoitered, touched, monitored and replayed, in part supplementing their carbon-based predecessors. Surreptitiously, almost insidiously, machines integrate themselves in our lives.[16]

Howard Bloom, an historian and evolutionist who, in an earlier vocational incarnation as a music agent, promoted the likes of Peter Gabriel, Bette Midler and Michael Jackson, projects the future of virtual reality in the following way. He asks us to imagine ourselves as a 70-year-old man, barely able to shuffle from the bedroom to the bathroom, logging on to a virtual dating system:

> You embark upon an adventure in a 360-degree synthesized computer-generated world of sight, sound, motion and touch. In virtual reality you look for a SWF under the age of 25 (or 18 if the scenario selected is medieval) who enjoys long walks into meadows filled with daisies. You shortly find her and she looks like Michelle Pfeiffer, the sexy movie star, except that she is more attractive. She is equally impressed with you.
>
> Why? Because on a digital date, you are no longer old, wrinkled and a heavy consumer of adult diapers. You look like...Arnold Schwarzenegger...You and the Michelle Pfeiffer look-alike take long walks in the daisy-filled fields of a springtime countryside and tell each other the secrets of your souls. Soon, you are in a mixed state of deep infatuation and unstoppable lust...What you don't know is that the person who has adopted this digital pose is actually 300 pounds, has never been able to walk within 30 feet of a chocolate bar without promptly adding it to her metabolic system, has 13 cats and is kept by her unfortunate bulk from cleaning their boxes more than once every six months. But buried under all that adipose tissue and the sweet perfume of feline urine, she has a delightful personality, and thanks to virtual communication, you are the 14th man she has driven to sexual bliss this month... On the other hand, she is only the second woman you've managed to conquer since

September....[such digital facades will] probably become a reality just in time for our decrepitude as senior citizens.[17]

"The computer's allure is more than utilitarian or aesthetic; it is erotic...our affair with information machines announces a symbiotic relationship...a mental marriage to technology...we feel augmented and empowered. Our hearts beat in the machines."[18] As we progress into the next millennium, individuals may become social cells and computers parts of neural networks in bastard new offspring: cyborg superorganisms.[19] Signaling that we are performing favorably, pleasure can increase the likelihood of repeating technological interactions. But it would be a mistake to consider our libidinous attachments to machines, our fascination with technology, as unnatural: rather, such attachments, incorporated into the great sexual cycles of life, increase the flow of information, the access to and recycling and degrading of energy. If orgasm is the experience of a wave of pleasure passing through the individual body, related through reproduction to gradient reduction, so the pleasure of electronically linking to distant others expands the gradient-reducing and reproductive activities of collective bodies. Eros, in other words, may metamorphose, losing its function in us as individuals but attaining a new one in future societies.

The human spirit is destined ever more frequently to discover itself extended into the realm of electromagnetic radiation. The most orgiastic prognostication in all science fiction is the repeatedly dreamed evolution of humankind into a species of pure energy. We must be on guard against the temptation to indulge in the millennial fervor-religious fantasy that we can completely escape our own bodies to exist in cyberspace or as bits of data in some storage medium.[20] Yet, certainly we have already expanded our activities across the electromagnetic spectrum. As hunter-gatherers our bodies absorbed light and emitted infrared radiation. Now through TV, x-ray and radioastronomy we broadcast into and receive complex messages from outer space across a far broader spectrum. And our videotape, recording and computer technologies allow us to store images and thoughts with greater fidelity over longer periods of time.

Despite the scope and apparent human uniqueness of our technologically mediated interconnectivity, densely packed societies evolving into organisms are not new. There has always been strength in numbers. Five hundred million years ago aggregated marine cells exported calcium ions to make calcium phosphate shells. Like a construction team laying iron girders, some later made internal calcium phosphate skeletons. Reproducing, cells organized into organisms. Their technology was in a way more advanced than ours, since the calcium they used had been a toxin which they opportunistically transformed into a building material.

Over billions of years, as life has become increasingly global, it has incorporated ever more chemical elements into its disperse bodies. Now life employs silicon—the abundant element of sand and sandstone—in its global communication. After oxygen silicon is the most common element in Earth's crust. But instead of using silicon to make bodies, life uses it to produce hyperbrains. Computers calculate using silicon instead, neurons to develop their connections. Semiconductors seem to supplant neurotransmitters. But, in fact, the cryogenically produced wafers, integrated circuits and other computer components do not supplant—they augment. They are becoming part of life's flesh. A single optical fiber the size of a human hair now transmits a trillion bits per second. This is fast enough to transmit every issue of the *Wall Street Journal* ever printed in less than one second.[21] Life, always growing, changing and exchanging matter with its environment, is incorporating old materials—silica derived from quartz and sand—into the latest biospheric organization. Exposed to the pressures of its own rapid growth, life long ago expanded from water to land and from land to air. Now life, on the brink of traversing outer space, delves into inner space. Life organizes atoms and electrons to form cyber-superorgan-

isms with living tendrils far beyond the Medusa and the chimeras of Greek mythology.

The technological interdependence of today, we predict, will alter dramatically the sex lives of our descendants. When we make love we go back in time. We produce bodily fluids, semen and vaginal lubricant reminiscent of the aquatic ancestral cellular environment. In ovulation, menstruation and ejaculation, we shed our protoctist like eggs, our uterine lining and our sperm. Two thousand million years ago, neither plant nor fungus nor animal dwelled on Earth's surface. Over a billion years before that there were only bacteria. Then, cells with nuclei evolved and went on to form the sexually reproducing protoctist ancestors to sexually reproducing plants, fungi and animals. [PLATE 58] But for all their body fanciness, complex evolution and elaboration, animals, plants and fungi still return each generation to the single- or few-celled state.[22] The mating of sexy nuclei still occurs in warm and wet tissues harboring body fluids. Cybersex, allowing on-line electronic coupling, may supersede this messy ancient imperative for an increasing number of individuals but, for the human species to continue, so must the wet ancient cycle persist in at least some.

The thrill of e-mail, instant messages and on-line chat rooms is the thrill of connection, of linking, of communication. With interconnection, conversational intercourse becomes more important and sexual intercourse less so. On-line desire facilitates face-to-computer interaction and lightning-fast intimacy among myriad partners as much as face-to-face romance or premarital skills. On-line technology gives humanity the potential to rearrange itself according to information not location; it makes possible stratifications and specializations as millions, then tens of millions, of people at widespread locations reorganize themselves on the planetary surface according to information flow.

Writer Gregory Stock has suggested "metaman."[23] Russian geochemist, Vladimir Vernadsky, and French paleontologist-theologian,

PLATE 58

The nubile sexuality of flowers has its evolutionary origin in leaves. Opening bud and young leaves of horse chestnut, Aesculus hippocastanum.

[Lois Brynes]

Teilhard de Chardin, wrote earlier in this century of "the noösphere."[24] We have spoken of Gaian-produced biospheres and a "bionic wave."[25] But there is, as yet, no accepted name for increasingly integrated, evolution-driven networks of interdependent humans and machines. The media speaks of the information superhighway. Yet such a highway leads nowhere but to itself and further connectedness between human beings. Far from being a travel route, the new information "superhighway" reduces the need for actual locomotion.[26] From an evolutionary standpoint the increasingly filigreed global speed-of-light communications system—motivated by the human desire, often sexual, to be connected—is better described as a new neural and perceptual infrastructure, the distributed intelligence of a multispecies developing global being. We need no more be aware of its intelligence than the constituent light-stimulated rod cells and their retinal neurons are aware of our larger selfhood and visual intelligence.

Is human sexual reproduction in a state of demise? Does not animal cloning, combined with birth control, mirror on the social level the sexless reproduction of our tissue cells, on the one hand, and of imposed limits of their once-unfettered growth on the other hand? *Stentor coeruleus*, a blue ciliate found in fresh water ponds and lakes, occasionally attempts mating but, when it does, both partners always die within less than a week.[27] Although the urge to merge occasionally resurfaces in these protoctists, sexual conjugations in *Stentor* are selected against, an evolutionary dead end. Although occasionally indulging, this organism, of course, does not require sex to reproduce. Indeed it must forego all sex in order to reproduce! So far, as a species, people do need sex for reproduction. But a day may come, and it may not be so very long from now, when sex and reproduction, in humans, will become permanently unhinged. Yet however unique it appears to us, the trend toward density-dependent reproductive moderation, including the decoupling of sex from reproduction, is nothing new. It has occurred many times in the history of evolution, especially in the transition from societies of cells and communities of microbes to large organisms.

We like to think of ourselves as special, higher than the rest of life, distinguished by our technology. But life, in fact, has always been "technological." Bacteria, protoctists and nonhuman animals incorporate external structural materials into their bodies and do construct dwellings. Our machines, like ourselves, are part of the sprawling nexus of planetary life that arose, leaving traces in the fossil record, well over three billion years ago. All organisms feel. All are alive. All can choose, even if on a minute scale, to alter their surroundings. Furthermore, all are connected—through the air, the water and the soil. Technology comes from life, not the other way round.

Each of us is a component in the urbanized human superorganism. People, united by powerlines, plumbing, sewage tunnels, copper wires, gas and air ducts, fiber-optic cables and global markets, act together. We are in no sense independent of each other nor of the biosphere. Together we ingest not only food but coal, iron, silicon, and oil: this last at a rate of over 70 million barrels per day. Networked by nation, culture and language we maintain our dwellings which include cities, roadways and fleets of nuclear submarines.

All organisms visible to the unaided eye are superorganisms. The most familiar example is you, a temperatur-regulatingmass of billions of cells stable at $37.5°$C ($98.6°$F). Bacteria, larger microbes, social insects, and even rodents construct living units of members that exist only in the aggregate. Now we humans are crossing the threshold from society to groups so interdependent we cannot survive unlinked to each other or unwired to our machines.

We require agriculture (including its tractors and trucks), medicine (including its hospitals and diagnostic scanners), industry (with its rigs and tankers) to survive in our current urban numbers. Yesterday's luxury becomes today's necessity. Any individual out of work or isolated from our community is more likely to be stressed, depressed or even fatally ill. Wiring—connecting up of members of societies into superorganisms—has been going on since the aggregation of cells into "individuals" on the early Earth. Once entirely dispensable as mere pleasure vehicles, automobiles are now essential for the working life and sustenance of millions of people. Once mere mathematical oddities, computers increasingly become integrated into our survival systems. Telephones that interlink us can spell the difference between life and death. The electronic conduits fanning out across Earth represent a new human chapter in the ancient evolutionary story of mergers and fusions of living beings. Reducing their totipotency and slowing down to form collectives more powerful than any single member, former confederacies become individuals at larger levels of organization. Sex in its many guises with its violent and variegated history has been essential in the past for human reproduction. Now these same sexual passions become crucial for new purposes as humans in society integrate further into cyborg superorganisms.

endnotes

Chapter 1
[PAGES 14-47]
A UNIVERSE IN HEAT : SEXUAL ENERGY

Epigraph: Redman, Alvin (ed.), 1959. *The Wit and Humor of Oscar Wilde* (from *Lady Windermere's Fan*), Dover: New York, p. 170.

1 All life can be assigned to one of five kinds (kingdoms or great groups)—the (1) bacteria, (2) protoctists and (3) fungi in all three of which sex is usually not required for reproduction—and the embryo-forming (4) plants and (5) animals that do usually require sex to develop and reproduce.

2 Despite lexical tradition, we employ the word gender throughout this work, in accord with the modern convention, to indicate not only verbal differences in sentences but physical differences that lead to mating in members of a species.

3 Vernadsky, W.I. (1944) "Problems of Biogeochemistry, II, The Fundamental Matter-Energy Difference between the Living and the Inert Natural Bodies of the Biosphere," (George Vernadsky tr.; G.E. Hutchinson ed.) In *Transactions of the Connecticut Academy of Arts and Sciences*, New Haven: Yale University Press, p. 489.

4 In the increasingly accepted five-kingdom classification scheme, protoctists are a miscellany of "large" microorganisms and their descendants ranging from tiny single-celled amebas and algae to large slime molds and giant kelp. Although some protoctists are plant-like and others resemble animals or fungi in their nutritional modes, members of this kingdom of water dwellers are neither plants nor animals, both of which grow from embryos. Nor are they fungi, land beings which grow from fungal spores (without forming embryos or the moving cell whips known as undulipodia). Because they have membrane-bounded nuclei (and for many other reasons), protoctists are not bacteria either. Protoctists are nucleated organisms that evolved from bacterial mergers. They do not belong to the other three living kingdoms (plants, animals, fungi) that also have merged bacteria for ancestors. For a professional look at this vast and still poorly known group see Margulis, L., Corliss, J.O., Melkonian, M., and Chapman, D.J., eds. (1990) *Handbook of Protoctista: the Structure, Cultivation, Habitats, and Life Histories of the Eukaryotic Microorganisms and their Descendants Exclusive of Animals, Plants and Fungi: A guide to the algae, ciliates, foraminifera, sporozoa, water molds, slime molds, and the other protoctists*, Jones and Bartlett Publishers, Inc.: Sudbury, Ma. For definitions and short explanations, see Margulis, L., McKhann, H. I., and Olendzenski, L., eds., S. Hiebert, editorial coordinator (1993) *Illustrated Glossary of Protoctista*, Jones and Bartlett Publishers, Inc., Sudbury, Ma.

5 Odum, Eugene (1953) *Fundamentals of Ecology*, Saunders: Philadelphia, quoted in Capra, Fritjof (1996) *Web of Life*, Anchor Books: New York, p. 176.

6 Morowitz, Harold (1987) *Cosmic Joy and Local Pain: Musings of a Mystic Scientist*, Charles Scribner's Sons: New York, pp. 92-93.

7 We do understand that the explanatory power of the Second Law can and has been occasionally overextended. A theory which explains everything, explains nothing. We remain convinced, however, that the Second Law, with its implicit recognition of one-way time, is crucial for understanding the physical backdrop to life's origin and evolution.

8 The Cambridge lectures were refashioned in Schrödinger, E. (1944) *What is Life?*, Cambridge

University Press: Cambridge. See also Margulis, L. and Sagan, D. (1995) *What is Life?*, Nevraumont/Simon and Schuster, New York.

9 The equations, however, are imperfect because they ignore friction, a thermodynamic phenomenon which in real life accrues only in one temporal direction.

For a technical treatment of the importance of energy flow, see Morowitz, Harold, J. (1968) *Energy Flow in Biology: Biological Organization as a Problem in Thermal Physics*, Ox Bow Press: Woodbridge, Connecticut.

For more on true time's alleged symmetry, see Price, Huw (1996) *Time's Arrow and Archimedes' Point: New Directions for the Physics of Time*, Oxford University Press: New York.

For more on mechanisms of global cooling, see Lovelock, James (1988) *Ages of Gaia*, W.W. Norton: New York, and Westbroek, Peter (1991) *Life as a Geological Force*, W.W. Norton: New York.

For more on gradient breaking, see Schneider, Eric D., and Kay, James J., (1995) "The Thermodynamics of Complexity in Biology," In *What is Life? The Next Fifty Years: Speculations on the Future of Biology*, Michael P. Murphy and Luke A.J. O'Neil, eds., Cambridge University Press: Cambridge, UK, pp. 161-173.

Chapter 2
[PAGES 48-83]
HOT AND BOTHERED: SEXUAL BEGINNINGS

Epigraph: Blake, William (1943) *The Poetical Works of William Blake*, John Sampson ed., Humphrey Milford, Oxford University Press: London, p. 133.

1 Sonea, Sorin and Panniset, Maurice (1983) *The New Bacteriology*, Jones and Bartlett: Sudbury, Ma.

2 One often reads about "bacterial chromosomes," but this is, strictly speaking, a misnomer. The bacterial genetic structure, called a genophore (when visible with the electron microscope) or chromoneme (when deduced from genetic experiments), does not have the histone proteins or tightly wound packaging (nucleosomes) composing the chromosomes (always more than two per cell) inside all eukaryotes.

3 The relationship of crisis to more complex forms of sex—that is, the meiotic sex of eukaryotes, some of which form spores sexually only as a response to extreme environmental stress—is discussed in Chapter 3.

4 We discussed what life is in Margulis, Lynn and Sagan, Dorion (1995) *What is Life?*, Nevraumont/Simon and Schuster. Since then, it has become even more apparent to us that life is one of a number of energy-degrading, so-called dissipative processes in the universe. See also, Schneider, Eric and Sagan, Dorion (1998) *Into the Cool: The New Thermodynamics of Creative Destruction*, Henry Holt and Company, New York (in preparation).

5 For a popular exposition of the importance of symbiosis in evolution, see Margulis, L. and Sagan, D. (1997) *Microcosmos: Four Billion Years of Microbial Evolution*, University of California Press: Berkeley. For technical details, consult Margulis

(1993) *Symbiosis in Cell Evolution* (2nd edition), W.H. Freeman Co: New York.

Chapter 3
[PAGES 84-123]
CANNIBALS AND OTHER VIRGINS: FUSION SEX

Epigraph: Vidal, Gore (1995) *Palimpsest: A Memoir*, Random House: New York, p. 23.

1 In zygotic meiosis, typical of fungi, after mating, the zygote undergoes immediate meiosis. In sporogenic meiosis, plant spores (not gametes) are formed from the diploid parent by meiosis. In gametic meiosis, typical of animals, meiosis occurs only to form gametes (eggs and sperm).

2 Most biological texts use the term "daughter cells." However, such terminology is at best confusing (and at worst downright wrong), since there is no necessary gender to the parent or the offspring cells. Needless to say, the problem is even worse when it is introduced into discussions of the origins of sex and gender!

3 Meiosis never evolved in single-cell amebas but it appeared in their multicellular descendants, the cellular slime molds. Meiosis evolved independently in ciliates, hypermastigotes, red algae, actinopods, foraminifera and perhaps other protoctists. Specific lineages of sexual protoctists are probably ancestors to plants (i.e., chlorophyte algae), fungi (i.e., chytrid protoctists) and animals (i.e., perhaps choanomastigotes). See **Appendix**.

4 Fungal "bodies" are masses of single (haploid) thready cells in support of their propagules—either meiotically (mushroom, morel) or mitotically (spore fuzz) produced. Plants grew bodies in the single state (for example, the male and female haploid moss gametophyte plants) that ended with mitotically produced gametes. The gametes fuse and form the doubled state. The long skinny moss-brown tower thread is called the sporophyte. Inside the tower's capsule the cells do meiotic division and end diploidy by meiotic spore formation. Many fungi persist in a sexually produced state where unfused (therefore haploid) nuclei float about in cytoplasm-fused threads. This haploid mass of fused cytoplasm with unfused nuclei is called the dikaryon.

5 Phyla of slime molds are distinguished by their fundamental biology: all cycle between single-celled individuals and multicelled blobs that to our eyes look like larger individuals. The Acrasiomycotes form amebas that after feeding swarm to make multicellular stalks that convert to walled spores. The Dictyostelids form amebas that after feeding swarm to make a shmoo-type migrating "slug" that travels as a single multicellular mass. Only later do they settle down to make multicellular stalks that convert to walled cells. Neither Acrasiomycotes nor Dictyosteliomycotes ever form undulipodiated cells. The life histories of the plasmodial slime molds, the Myxomycotes, are far more complex. Individual myxomycote cells alternate between amebas, amebomastigotes, fusing mastigotes, fusing amebas as well as huge multinucleate plasmodia that make structures that look like plants.

6 Franks, N.R. (1989) "Army Ants: A Collective Intelligence" *American Scientist*, 77:139-45. Also, see Wilson, E.O. (1987) "Causes of Ecological Success: The Case of the Ants," *J. Anim. Ecol* 56: 1-9. For a more general treatment see, Corning, Peter (1996) Holistic Darwinism: Group Selection and the Bioeconomics of Evolution, 19th Annual Meeting European Sociobiological Society, July 22-25. And see Bloom, Howard (1995) *The Lucifer Principle: A Scientific Expedition into the Forces of History*. The Atlantic Monthly Press: New York.

7 When we say animals are "higher" we really mean that they are "more like us." When we call animals like us "higher" we may be intuitively referring to their distance from thermodynamic equilibrium. RNA, a bacterium, merged bacteria, nucleated cells, fused nucleated cells, and sexual bodies formed of specifically growing fused nucleated cells are each a further step away from the "ground state" of inert (non-living) matter.

8 The point that uniparental animals really do not belong to species because they have no opportunity to intrabreed as part of a larger population is made in "Sex and the Order of Nature," in Wicken, Jeffrey S. (1987) *Evolution, Thermodynamics, and Information: Extending the Darwinian Program.* Oxford University Press: New York, pp. 212-219. We believe that species, in practice, are defined by their morphological distinctions, a "morphospecies" concept. Any individuals that have the same number and same types of integrated former free-living organisms comprising it belong to the same species. The criterion of mating compatibility to define species, applicable to many plants, fungi, protoctists and animals, is limited to certain taxa only: e.g., most birds, mammals and insects. This species definition is not useful for life in general. Given the morphological species definition, uniparental animals, like whiptail lizards, are easily classified from kingdom to species.

9 Carrol, Lewis (1871) *Through the Looking-Glass and What Alice Found There*, Macmillan: London. Cited Ridley, Matt (1994) *The Red Queen, Sex and the Evolution of Human Nature*, Macmillan Publishing Company: New York, p. 64.

10 Gonick, Larry (1995) "Science Classics," in *Discover*, December, pp. 108-109. See also, Margulis, L. and Sagan, D. (1990) *Origins of Sex*, Yale University Press, New Haven, CT; Sagan, D. and Margulis, L. (1985) "The Riddle of Sex," *The Science Teacher* 52:16-22.

Chapter 4
[PAGES 124-147]
**THE KISS OF DEATH:
SEXUALITY AND MORTALITY**

Epigraph: Rimbaud, Arthur (1957) *Illuminations, and Other Prose Poems*, New Directions: New York, p. 147.

1 Dobson, John L. (1995) "The Equations of Maya," In *Cosmic Beginnings and Human Ends: Where Science and Religion Meet*, (Clifford N. Matthews and Roy Abraham Varghese, eds.) Open Court: Chicago, p. 272.

2 Jantsch, Erich, (1983) *The Self-Organizing Universe*, Pergamon Press: New York, p. 16.

3 Again, the term species may be misleading since in these populations organisms of complementary genders do not necessarily breed with each other to produce offspring; they do not share a common gene pool when they reproduce uniparentally. Although they cannot be considered species in the traditional zoological sense of the word, they certainly are recognized as a "morphospecies." For more examples, see White, M.J.D. (1961) "The Cytology of Parthenogenesis," In *The Chromosomes*, John Wiley and Sons; New York, pp. 127-137.

4 Holdrege, Craig (1996) *Genetics and the Manipulation of Life*, Lindisfarne Press: Hudson, New York.

5 Ibid, pp. 110-11.

6 We know this concept of relative delay (of centromere versus chromosome reproduction or visa versa) is difficult to understand—and that we have not explained it here. For the necessarily technical details, including explanation of centromeres in karyotypic fissioning, see Margulis, L. and Sagan, D. (1990) *Origins of Sex*, Yale University Press: New Haven; and especially Margulis, L. (1993) *Symbiosis in Cell Evolution*, 2nd

ed. Or see shorter description in "The Riddle of Sex," chapter 21, of our 1997 *Slanted Truths*, Springer Verlag: New York, pp. 283-94. Note: this chapter's discussion of cell death is partially based on and gratefully acknowledges William, R. Clark's *Sex and the Origins of Death* (Oxford University Press: New York and Oxford, 1996).

Chapter 5

[PAGES 148-199]

STRANGE ATTRACTIONS: SEX AND PERCEPTION

Epigraphs: Darwin, Erasmus [Charles' grandfather] (1794) *Zoonomia: The Laws of Organic Life*, Vol. 1., J. Johnson, London, pp. 503 and 519. Also, Wilde, Oscar, 1959, *The Wit and Humor of Oscar Wilde* (from *A Woman of No Importance*), Alvin Redman ed., Dover Publications: New York, p. 105.

1 Schneider, Eric, D. and Kay, James, J., 1995, "Order from Disorder: The Thermodynamics of Complexity in Biology," in Murphy Michael, P. and O'Neil, Luke A.J., eds., *What is Life? The Next Fifty Years: Speculations on the Future of Biology*, Cambridge University Press: Cambridge, pp. 168-170.

2 Abram, David (1996) *The Spell of the Sensuous*, Pantheon Books: New York.

3 Darwin, Charles (1874) *The Descent of Man, and Selection in Relation to Sex*, Murray: London, p. 257.

4 Andersson, M. (1982) "Female Choice Selects for Extreme Tail Length in a Widow Bird," *Nature* 299: 818-820.

5 Wallace, A. R. (1901) *Darwinism*, 3rd ed, Macmillan: London, p. 273.

6 In Darwin, F. and Seward, A.C., eds. (1903) *More Letters of Charles Darwin: A Record of His Work in a Series of Hitherto Unpublished Letters*, John Murray: London, pp. 62-63.

7 Hamilton, H.D. and Zuk, M. (1982) "Heritable True fitness and Bright Birds: A Role for Parasites?," *Science* 218:384-387.

8 Eberhard, William G. (1985) *Sexual Selection and Animal Genitalia*, Harvard University Press: Cambridge, p. 83.

9 A Hobson's Choice is not a choice at all—it is a response to a clever magician's subtle coercion. The interesting "forced" nature of female sexual choice raises the larger issue of choice in general. As is well known, free will does not jibe well with the prevailing scientific world view in which everything, including life, is the largely deterministic result of past interactions. But the feeling of being able to choose to read this sentence to its end or put this book down—and to make countless other decisions—is immediate and strong. The "folly of choice" (this phrase we owe to Robin Kolnicki) within the modern scientific world view forces us to make one of two conclusions: either free will exists, and science must somehow make room for it, or it is an illusion. Could it be, we wonder, that the feeling of being able to choose is a mental "short cut" allowing us to make sense of decisions whose sheer number would otherwise overwhelm us? And might not determined, automaton-like organisms, with an illusion of free will, be more likely to survive than biochemical robots without such dementia?

10 Eberhard, William G. (1985) *Sexual Selection and Animal Genitalia*, Harvard University Press: Cambridge, p.71.

11 Smith, N.G. (1967) "Visual Isolation by Gulls," *Scientific American*, 217 (4): 94-102.

12 There is the philosophical objection, however, that from a vantage point of radical doubt, nothing really safeguards an evolutionary epistemology: if a belief about nature helps us survive, it will be adhered to whether not it corresponds to some external reality. For example, the belief in spiteful wind and rain gods persist if it encourages greater care of crops by tribespeople.

13 Shephard, Roger, N. (1990) *Mind Sights: Original Visual Illusions, Ambiguities, and Other Anomalies, with a Commentary on the Play of Mind in Perception and Art*, W.H. Freeman, New York, p. 4.

14 Delbruck, Max (1985) "An Essay on Evolutionary Epistemology" in *Mind from Matter*, eds. Gunther S. Stent and E. Peter Fischer, Blackwell Scientific Publications, Oxford, UK.

15 Angier, Natalie (1996) "Illuminating How Bodies are Built for Sociability," *The New York Times*, April 30, 1996, pp. CI and CII. Also, see Ackerman, Diane (1994) *A Natural History of Love*, Vintage Books: New York, pp. 166-167.

16 Liebowitz, Michael (1983) *The Chemistry of Love*, Boston: Little Brown.

17 Ackerman, Diane (1994) *A Natural History of Love*, Vintage Books: New York, pp. 164-166.

18 Trivers, R.L. (1985) *Social Evolution*, Benjamin Cummings: Menlo Park, California.

19 Smith, Robert, ed. (1984) *Sperm Competition and the Evolution of Animal Mating Systems*, Academic Press: Orlando, Florida. Also, see Baker, R.R. and Bellis, M.A. (1994) *Human Sperm Competition: Copulation, Masturbation and Infidelity*, Chapman and Hall: London.

20 The old taxonomic name for chimps, *Pan satyrus*, refers to the myth of apes as lustful satyrs. The English word, gorilla, is derived from the Greek name for a tribe of hairy African women would have been a better name for bonobos than *Pan paniscus*, according to Emory ethnologist Frans B. M. de Waal. See de Waal, Frans B. M. (1995) "Bonobo Sex and Society," *Scientific American*, March: 82-88.

21 The "right to lifers'" refusal to allow women to receive abortions, in light of the evolution of breeding systems, is an institutionalized form of male ancient possessive jealousy. We interpret this anti-abortion activity as, in part, an attempt by "dominant" males' to impede impregnation by other males of women who are already impregnated. Reproductive control of women's bodies ensures that the long-evolved female choice that Darwin avidly chronicled in the animal realm continues to apply in the human realm, as females, by choosing with whom they breed, continue to shape human evolution.

22 Frank, Laurence, G., Weldele, Mary, L., and Glickman, Stephen, M. (1995) "Masculinization Costs in Hyenas," *Nature* 377:6550. In "Sex and the Spotted Hyena," a colloquium delivered February 14, 1996, in the Neuroscience and Behavior Program at the University of Massachusetts at Amherst, Glickman reported on research that found that hyena ovaries produce a little estrogen and a lot of androstenedione, the chemical precursor to both testosterone and estradiol, an estrogen. Androgens that normally circulate in the mother's blood are protected by binding proteins that neutralize the potentially sex-changing chemicals in human and other female mammalian fetuses. In hyenas, however, the placenta synthesizes testosterone from androstenedione produced by the ovaries and circulating in the mother's blood; the placenta manufactures testosterone near the developing female embryo. An enzyme called aromatase converts testosterone to estradiol, the hormone responsible for ensuring the femaleness of the infant. Since the hyena studies were conducted, three genetically

female humans in San Francisco who show secondary sexual features of men have been diagnosed with an aromatase deficiency.

Chapter 6
[PAGES 200-228]
COME TOGETHER: THE FUTURE OF SEX

Epigraph The quote by geochemist Robert Garrels was a personal communication to Dorion Sagan at the cafeteria at the University of California, San Jose in 1984.

1 The terms confirmation and novelty to describe these two crucial poles of life's dynamic are a coinage of astrophysicist Erich Jantsch. See Jantsch, Erich (1983) *The Self-Organizing Universe*, Pergamon Press: New York.

2 Comparisons of genetically identical water fleas, locusts, rabbits and owl monkeys have all been shown to exhibit neural variability as a result of environmental exposure and individual development patterns. See Edelman, Gerald M. (1992) *Bright Air, Brilliant Fire: On the Matter of the Mind*, Basic Books: New York, p. 26.

3 Our relationship with water underscores our quasiconvergence with water-dwelling animals that evolved from land-inhabiting ancestors. Several distinct features of humans, including our flexible spine, marine-animal-like ability to weep copiously, fondness for baths and beaches, water-friendly relative hairlessness, gag reflex, native ability of our infants to learn quickly to swim, finlike hands and occasional webbed feet, have suggested to some that, unlike other apes, we underwent a formative stage of evolution in the water. The people in Kurt Vonnegut's novel *Galapagos* convergently evolve. A few rich tourists and celebrities on a remote isle safe from World War III and a pandemic AIDS-like virus eventually develop blubber and smaller pointy heads, barely splashing when they dive into the water and incapable of remembering their mothers after the age of three; Vonnegut pictures these aquatic human descendants lying out sunbathing and reminds the reader that the only trait firmly tying them to their human ancestors is that, when one of them farts while lying out on the beach, the others predictably burst out into laughter. In a sort of half-hewn version of Vonnegut's fictional scenario, the aquatic theory suggests that, like the ancestors of walruses, sea lions, seals, dolphins and whales, human mammal ancestors returned to the water. Unlike these organisms, however, we then readapted ourselves to the land. The theory that we underwent a crucial period of evolution near and in the water, perhaps during the Pliocene, was first put forth by marine biologist Sir Alister Hardy when he tried to account for the layer of subcutaneous fat, mentioned by an anatomist, possessed by humans but not our closest relatives, the chimpanzees. "Hardy's vision...of an early primate being driven onto offshore islands on what was then the east coast of Africa...of a 'tropical penguin'...has been criticized by the more entrenched traditionalists, who point out that it is entirely conjectural and there is no single shred of direct evidence to support it. What they fail to admit is that their own 'savannah hunting theory' is equally circumstantial." See Morris, Desmond (1994) *The Human Animal: A Personal View of the Human Species*, Crown Publishers, Inc.: New York, pp. 53-61. The importance in the human diet of essential fatty acids, obtained readily from fish, might also support the Hardy notion of aquatic ancestry.

4 Butler, Samuel (1924) *Unconscious Memory*, Vol. 6 of *The Shrewsbury Edition of the Works of Samuel Butler*, Jonathan Cape: London, p. 57. For more on the evolutionary transition from societies to organisms, see the following references: Sagan, D. (1992) "Metametazoa: Biology and Multiplici-

ty" In *Incorporations* (Zone 6; *Fragments for a History of the Human Body*) Jonathan Crary and Sanford Kwinter, eds., Zone: New York, pp. 362-385; Sagan, D. (1997) "What Narcissus Saw: The Oceanic "I"/Eye." In *Slanted Truths: Essays on Symbiosis, Gaia and Evolution*, Springer Verlag, New York, and Sagan, D. (1990) *Biospheres*, Bantam Books, New York. See our exposition of Samuel Butler's theories in Chapter 9 of Margulis, L. and Sagan, D. (1995) *What is Life?* Nevraumont/Simon and Schuster: New York.

5   Sherman, Paul W., Jarvis, Jennifer U. M., and Alexander, Richard, D. (1991) *The Biology of the Naked Mole-Rat: Monographs in Behavior and Ecology*, Princeton University Press: Princeton, NJ.

6   Although we find it difficult to conceive of a body without a head or a nation without a king, prime minister, president, chief or other head-of-state, organic stratification does not include a head and its implied top-down structure of a hierarchy. Holons are entities composed of smaller holons as communities are composed of individuals. Taken together holons form "holarchies." Holarchies refer to organizations with many different centers of power, many "heads." Nonetheless, the central importance of a head to our notion of body alerts us that focusing power and communication in a central area is an oft-tread route in biological reorganization.

7   Freud, Sigmund (1955) *Civilization and its Discontents*, Joan Riviere, tr., The Hogarth Press Ltd.: London, pp. 76-77.

8   Colborn, Theo, Dumanoski, Dianne, and Myers, John Peterson (1996) *Our Stolen Future: Are We Threatening Our Fertility, Intelligence, and Survival?—A Scientific Detective Study*, Dutton: London.

9   Gimeno, Sylvia, Gerritsen, Anton, and Bowmer, Tim (1996) "Feminization of Male Carp," *Nature* 384(21):221-222.

10 Fausto-Sterling, Anne (1993) "The Five Sexes: Why Male and Female are Not Enough," *The Sciences*, March/April, pp. 20-24.

11 Terman, Richard, C. (1984) "Sexual Maturation of Male and Female White-Footed Mice (*Peromyscus leucopus noveboracensis*): Influence of Physical or Urine Contact with Adults," *Journal of Mammalogy*, 65(1):97-102.

12 B.R. Komisaruk and B. Whipple have further demonstrated the vagus pathway in rats by severing spinal cords, stimulating the rats' cervixes, and then observing pupil dilation and an increase in the animals' threshold to pain. They also removed sections of the spinal cord and found the same results. In 1990, Matthew J. Wayner at the University of Texas in San Antonio injected a tracer chemical into rat genitalia and observed that the tracer was taken up by the vagus nerve—"indicating that there was a pathway that circumnavigated the spinal cord." DeKoker, Brenda (1996) "Sex and the Spinal Cord: A new pathway for organism," *Scientific American* 275 (6), pp. 30-32.

13 Shaviro, Steven (1997) *Doom Patrols: A Theoretical Fiction About Postmodernism*, Serpent's Tail: New York and London, pp. 37-38.

14 Biologists distinguish between so-called "R" selection, pertaining to species like oak trees, frogs and turtles which typically produce many offspring few of which survive, and "k" selection, which describes organisms like orchids, kangaroos and baboons that produce fewer offspring but devote more attention and resources to each, enhancing the relative rate of survival. Most fish and cockroaches are "R" selected whereas elephants and humans are "k" selected where the letters refer to the coefficients in a population equation. Indeed, the prognosticated loss of totipotency in humans is in a sense a logical extension of "k" selection.

15 Even the self-conscious ironic promulgation of being intelligent without having a place, a commensurate social function, a "life"—the "slackers" of "Generation X"—can be viewed as symptomatic of socio-reproductive reorganization.

16 The status of technology is an interesting futuristic question. Is it an extension of humanity that adds to our grandeur or, rather, something incubated by humanity that will eventually escape our control and overtake us? Are innocent appliances leading to a day of artificial intelligence and robots running amok, taking over their masters? Or is technology a kind of second skin and supplementary organ system that gives those who control it the advantages of a superman? Publishing in a New Zealand newspaper under different pseudonyms Samuel Butler argued both sides of the technology question in the 19th century. For more on Butler and machines, see Dyson, George B. (1997) *Darwin Among the Machines: The Evolution of Global Intelligence*, Addison-Wesley: Reading, Massachusetts. Also, Sagan, D. (1990) *Biospheres*, Bantam Books: New York, and the last chapter of our 1995 *What is Life?* Foreword by Niles Eldrege, Nevraumont/Simon and Schuster: New York.

17 Bloom, Howard (1995) "Love with the Proper Stranger," *Net Guide*, (February), pp. 1-2. See also, Turkle, Sherry (1995) *Life on the Screen: Identity in the Age of the Internet*, Simon and Schuster: New York.

18 Heim, Michael (1991) "The Erotic Ontology of Cyberspace," in Benedikt, Michael, ed. *Cyberspace: First Steps*, The MIT Press: Cambridge, Massachusetts. p. 61.

19 Haraway, Donna (1991) "A Cyborg Manifesto: Science, Technology, and Socialist-Feminism in the Late Twentieth Century" in *Simians, Cyborgs, and Women: The Reinvention of Nature*, New York: Routledge.

20 Dery, Mark (1996) *Escape Velocity: Cyberculture at the End of the Century*, Grove Press: New York.

21 Negroponte, Nicholas (1995) *Being Digital*, Alfred A. Knopf: New York, p. 23.

22 Animals, of course, form a haploid egg (a single cell) fertilized by one haploid (single cell) sperm, although many sperms attempt to penetrate. Fungi return to single cells or very few cells when they form propagules called conidia (haploids—single parent only needed) or ascospores or basidiospores (two parents needed but still haploid). The story of the return to single state plants is somewhat more complex—mating is wet and sweet but in flowering plants it occurs between the nucleus of elongated pollen grain and the nucleus of the female—the embryo sac fully surrounded by floral tissue. The generalization holds: all large sexual organisms return each generation to the single cell (sperm, egg, fungal spore) or its equivalent (multicelled fungal spore, pollen tube nucleus, embryo sac nucleus, etc.). See Margulis and Sagan (1994) *What Is Life?*, Nevraumont/Simon & Schuster: New York.

23 Stock, Gregory (1993) *Metaman: Humans, Machines, and the Birth of a Global Superorganism*, Bantam Press: London.

24 Grinevald, J. (1988) "A History of the Idea of the Biosphere," In P. Bunyard and E. Goldsmith (eds.) *Gaia: The Thesis, the Mechanisms and the Implications. Proceedings of the First Annual Camelford Conference on the Implications of the Gaia Hypothesis* Quintrell and Co.:Cornwall, UK. Reprinted as Grinevald, J. (1996) "Sketch for a History of the Biosphere," In P. Bunyard (ed.) *Gaia in Action: Science of the Living Earth*, Floris Books: Edinburgh, pp. 34-53.

25 Sagan, D (1990) *Biospheres*, Bantam Books: New York, pp. 6-7.

26 The "movement" away from mobility is reflected in the composition of the stocks making up the S&P 500 stock index. In 1960, there were 43 transportation, 24 auto and truck, and only 8 computers and telecommunications companies listed. By 1995, the computers and telecommunications group had risen to 37 while only 15 transport and 13 auto and truck companies remained.

27 For more on the peculiarities and technicalities of sex in the microworld see Margulis, L. and Sagan, D. (1990) *Origins of Sex* Yale University Press, New Haven, CT. For an opening into the expanding literature on the genetic control of these sexual peculiarities see Horgan, John (1997) "A mutant gene alters the sexual behavior of fruitflies," *Scientific American*, June, pp. 26-31.

appendix: SEX LIVES *Phyla and examples of genera*

BACTERIAL SEX (See Plates 8-11.)

SUPERKINGDOM PROKARYA (KINGDOM BACTERIA)

The genes in these organisms are loose (not contained with membrane-bounded nuclei). They lack cell structures (organelles) such as mitochondria or chloroplasts. They have exerted a more profound influence than other organisms on the biosphere for more than 3 billion years. Their modes of sexuality are summarized in Plate 9. In bacteria one partner (the recipient) undergoes genetic change (becomes a new organism) by having sex (by acquiring donor genes) from another bacterium. But because the number of organisms does not increase (no new organisms are generated), bacterial sex is not a reproductive process.

SUBKINGDOM ARCHAEA (ARCHAEBACTERIA)
Phylum EURYARCHAEOTA (methanogens, halophils) *Halobacter, Methanococcus*
Phylum CRENARCHAEOTA (thermoacidophils) *Pyrodictum, Thermoplasma*

SUBKINGDOM EUBACTERIA
Phylum PROTEOBACTERIA (purple bacteria) *Escheria, Pseudomonas*
Phylum SPIROCHAETAE (spirochetes) *Spirosymplokos, Treponema*
Phylum CYANOBACTERIA (oxygenic photosynthesizers) *Nostoc, Anabaena*
Phylum SAPROSPIRAE (fermenting gliders) *Bacteroides, Cytophaga*
Phylum CHLOROFLEXA (green nonsulfer phototrophs) *Chloroflexus*
Phylum CHLOROBIA (anoxygenic green sulfer bacteria) *Chlorobium*
Phylum APHRAGMABACTERIA (wall-less bacteria) *Mycoplasma, Spiroplasma*
Phylum ENDOSPORA (spore-forming rods) *Bacillus, Clostridium*
Phylum PIRELLULAE (stalked bacteria with protein walls) *Pirellula*
Phylum ACTINOBACTERIA (fungus-like bacteria) *Streptomyces*
Phylum DEINOCOCCI (radioresistant Gram-positive bacteria) *Deinococcus*
Phylum THERMOTOGAE (thermophilic fermenters) *Thermotoga*

The details of the sex lives in all the above organisms (Subkingdoms Archaea & Eubacteria) are not known.

SUPERKINGDOM EUKARYA

The Eukaryotes are made of larger, more complex cells with membrane-bonded nuclei; DNA is organized into chromosomes (at least two and sometimes over a thousand) inside these nuclei. All evolved from hypersex (See Plate 7), i.e., from ancient symbiotic consortia whose smaller constituents share common ancestors with different kinds of modern bacteria. Eukaryotic cell nuclei are either haploid or diploid, e.g. carry either one or two complete sets of chromosomes. The process of meiosis, which halves the chromosome number, insures that later fusion of sex cells (usually haploid) will restore diploidy in the zygote. Hence, eukaryotic sexuality involves meiosis, and is often coupled to reproduction (e.g., in plants and animals a new organism grows from the sexually fertilized egg).

KINGDOM PROTOCTISTA

All protoctists evolved from integrated permanent symbioses between different kinds of bacteria (e.g., green algae from archaebacteria+ eubacteria+ cyanobacteria permanent associations). Algae, protozoa, slime molds, water molds, giant kelp, slime nets and many others are included in protoctists. About 250,000 species are extant. Protoctists are the organisms in which mitotic cell reproduction and fertilization sex evolved. Protoctists do not form embryos or fungal spores. All are eukaryotic, i.e., all have membrane-bounded nuclei. Some lack common names.

KEY
+ = Presence of meiotic sex in at least one species
− = Absence of meiotic sex in all taxa studied to date
? = Sexuality may occur; not known for sure

Amitochondriates (Hypochondria, Archaezoa; lack mitochondria)
+   Phylum ARCHAEPROTISTA *Pelomyxa, Trichomonas, Giardia*
+   Phylum MICROSPORA *Glugea, Vairamorpha*

Ameboids
−   Phylum RHIZOPODA (shelled and naked amebas) *Ameba, Dictyostelium*
+   Phylum GRANULORETICULOSA (foraminiferans) *Globigerina, Rotaliella*
?   Phylum XENOPHYOPHORA (large deep-sea protoctists) *Galatheammina*
+   Phylum MYXOMYCOTA (slime molds) *Cercomonas, Physarum*

Alveolates (gas- or liquid-filled cavity formers)
+   Phylum DINOMASTIGOTA ("red tide" algae) *Gonyaulax, Gymnodinium*
+   Phylum CILIOPHORA (ciliates) *Gastrostylax, Paramecium*
+   Phylum APICOMPLEXA (cell-piercing parasites) *Eimeria, Toxoplasma*

Swimming mastigotes
−   Phylum HAPTOMONADA (limestone-forming plankton) *Emiliana, Prymnesium*
−   Phylum CRYPTOMONADA *Cryptomonas, Cyathomonas*
−   Phylum DISCOMITOCHONDRIA (protists with discoid mitochondrial cristae)
     Naegleria, Trypanosome, Euglena

Heterokonts (stramenopiles; bearers of two unequal undulipodia)
- Phylum CHRYSOMONADA (golden yellow algae) *Ochromonas, Synura*
- Phylum XANTHOPHYTA (yellow green algae) *Ophiocytium*
- Phylum EUSTIGMATOPHYTA (yellow green algae) *Vischeria*
+ Phylum BACILLARIOPHYTA (diatoms) *Coscinodiscus, Navicula*
+ Phylum PHAEOPHYTA (brown algae; kelp) *Ascophyllum, Fucus*
+ Phylum LABYRINTHULATA (slime nets) *Labyrinthula, Thrastochytrium*
+ Phylum PLASMODIOPHORA (multinucleate plant parasites) *Sorodiscus*
+ Phylum OOMYCOTA (egg molds) *Phytophthora, Saprolegnia*
- Phylum HYPHOCHYTRIDIOMYCOTA (water molds) *Anisolpidium, Canteriomyces*

Propagule-forming parasites
- Phylum HAPLOSPORA (unicellular spore-forming parasites) *Haplosporidium*
- Phylum PARAMYXA (multicellular spore-forming parasites) *Marteilia*
+ Phylum MYXOSPORA (cnidocyst-forming fish parasites) *Myxobolus*

Conjugating algae (complementary-gender seaweeds)
+ Phylum RHODOPHYTA (red algae) *Amphiroa, Polysiphonia*
+ Phylum GAMOPHYTA (CONJUGAPHYTA) (conjugating green algae) *Micrasterias*

Heliozoa and radiolaria (star-shaped test-forming plankton)
+ Phylum ACTINOPODA *Acanthocystis, Acantharia*

Ancestral phyla
+ Phylum CHLOROPHYTA (green algae; plant ancestors) *Acetabularia*
+ Phylum CHYTRIDIOMYCOTA (fungal ancestors) *Blastocladiella*
- Phylum ZOOMASTIGOTA (animal ancestors) *Jakoba, Acronema, Zelleriella*

KINGDOM ANIMALIA
All animals evolved from some (unknown) members of the Protoctista, most likely from members of Phylum Zoomastigota. They develop from the sexually-produced blastula. Groups in which apomixis (parthenogenesis)—secondary loss of biparentality; uniparental sex—is present in at least one species are marked with "+", if apomixis is likely but not proven "?". All other phyla practice biparental sexuality.

Sponges and mesozoa
? Phylum PLACOZOA *Trichoplax*
+ Phylum PORIFERA (sponges) *Gelliodes*
? Phylum RHOMBOZOA *Dicyema*
? Phylum ORTHONECTIDA *Orthonectis*

Radiates (animals with radial symmetry)

+ Phylum CNIDARIA (jellyfish, corals) *Craspedacusta, Hydra*
   Phylum CTENOPHORA (comb jellies) *Bolinopsis, Ctenoplana*

Aperitoneal worms (worms lacking coelom or body cavity)

   Phylum GNATHOSTOMULIDA *Problognathia*
+ Phylum PLATYHELMINTHES (flatworms) *Planaria, Procotyla*
+ Phylum GASTROTRICHA *Tetranchyroderma*
+ Phylum ROTIFERA (rotifers) *Brachionus*
   Phylum KINORHYNCHA *Echinoderes*
   Phylum LORICIFERA *Nanaloricus, Pliciloricus*
   Phylum ACANTHOCEPHALA *Leptorhynchoides, Macracanthorhynchus*
   Phylum NEMATODA (nematodes) *Rhabdias*
   Phylum NEMATOMORPHA *Gordius*
   Phylum PRIAPULA *Tubiluchus*
   Phylum ENTOPROCTA *Barentsia, Loxosoma*

Peritoneal worms (worms with coelom or body cavity)

   Phylum RHYNCHOCOELA (nemertines) *Prostoma*
   Phylum SIPUNCULA *Themiste*
   Phylum ECHIURA *Metabonellia*
   Phylum ANNELIDA (segmented worms) *Lumbricus, Hirudo*
   Phylum POGONOPHORA (beardworms) *Riftia*
+ Phylum MOLLUSCA (clams, snails, squid) *Ensis, Crepidula, Loligo*

Nonjointed foot-walkers

   Phylum TARDIGRADA (water bears) *Echiniscus*
   Phylum PENTASTOMA (tongueworms) *Linguatula*
   Phylum ONYCHOPHORA (velvet worms) *Speleoperipatus*

Arthropods (jointed foot-walkers)

   Phylum CHELICERATA (spiders, horseshoe crabs) *Araneus, Limulus*
+ Phylum MANDIBULATA (insects, centipedes)
   Phylum CRUSTACEA (crabs, lobsters, barnacles) *Balanus, Parthenope*

Lophophorates (bearers of lophophores, these are ciliated, horseshoe-shaped food-gathering organs)

   Phylum BRYOZOA *Plumatella, Symbion*
   Phylum PHORONIDA *Phoronis, Phoronopsis*
   Phylum BRACHIOPODA *Terebratulina*

Achordate deuterostomes (second mouth-formers lacking notochord, embryonic spinal cord precursor)

      Phylum ECHINODERMATA (starfish, sea urchins) *Aterias, Cucumaria*
      Phylum CHAETOGNATHA (arrow worms) *Bathybelos, Sagitta*
      Phylum HEMICHORDATA *Ptychodera, Saccoglossus*

Chordate deuterostomes (second mouth-formers with notochord)

      Phylum UROCHORDATA (tunicate sea squirts) *Didemnum*
      Phylum CEPHALOCHORDATA *Branchiostoma*
+   Phylum CRANIATA (fish, amphibians, reptiles, birds, mammals) *Homo*

KINGDOM FUNGI

Fungi evolved from protoctists, most likely from members of the Phylum Chytridiomycota. All develop from spores (in haploid stage) whether sexually produced or not. + = Apomixis: secondary loss of biparentality; uniparental sex present in at least one species. Most species indulge in biparental sex.

+   Phylum ZYGOMYCOTA (bread molds; fungus) *Mucor, Rhizopus*
+   Phylum BASIDIOMYCOTA (mushrooms) *Agaricus, Schizophyllum*
+   Phylum ASCOMYCOTA (fungi imperfecti, lichens, yeasts) *Neurospora*

KINGDOM PLANTAE

All evolved from some protoctists, most likely green algae (Phylum Chlorophyta). All develop from spores (in haploid stage) and from sexually produced embryos (in diploid stage). + = Apomixis: secondary loss of biparentality; uniparental sex present in at least one species.

      Phylum BRYOPHYTA (mosses) *Polytrichum, Takakia*
+   Phylum HEPATOPHYTA (liverworts) *Marchantia*
      Phylum ANTHOCEROPHYTA (hornworts) *Anthoceros*
      Phylum SPHENOPHYTA (horsetails) *Equisetum*
+   Phylum FILICINOPHYTA (ferns) *Osmunda, Polypodium*
      Phylum GINKGOPHYTA *Ginkgo*
      Phylum CONIFEROPHYTA (conifers) *Pinus, Taxus*
      Phylum GNETOPHYTA *Ephedra, Welwitschia*
      Phylum CYCADOPHYTA (sago palms) *Cycas, Zamia*
+   Phylum ANTHOPHYTA (flowering plants) *Aster, Avena, Oenothera*

glossary

Allelochemical Environmental hormone; a substance produced by an organism of one kind, released into the environment and responded to by a different kind of organism. Examples include air- or waterborne toxins, sex attractants, fruit ripeners and odiferous compounds made by flowers resembling putrefying meat that attract flies. Pheromones are any allelochemics that act on members of the same species whereas sex pheromones are those produced by one gender and responded to by the other. See *pheromone*.

Anisogamy Sex cells (gametes) requiring fertilization for continued development that differ from each other in size or form (e.g., egg and sperm of water molds or animals are anisogametes whereas the two equal gametes of *Chlamydomonas* are isogamous). Anisogamy occurs in animals and plants. See *isogamy*.

Apomixis (adj. apomictic) Condition of being formerly sexual. Single-parent meiosis or fertilization such that fertile two-parent matings (*mixis*) is bypassed, e.g. "virgin birth."

Apoptosis Programmed cell death; in contrast to *cytocide* (see below), the genetically encoded death of a cell which occurs during normal development.

Apoptotic bodies Fragments released by the dying cell; DNA and membrane-rich small structures that are products of apoptosis.

Assortative mating Mating or sexual selection that is not random wherein similar individuals mate more successfully (produce more offspring) relative to different (unlike) ones. Successful mating with relatives is assortative whereas productive mating with distant members of entirely different unrelated populations is disassortative.

Autogamy Self-fertilization; mating (fusion) of two whole cells or nuclei both derived from a single parent cell or nucleus.

Autopoiesis Self-maintenance; set of principles defining life and pertaining to membrane-bounded, self-limited, internally organized systems that dynamically maintain their identity in a changing environment. Autopoietic entities are able to replace and repair their constituent parts, ultimately at the expense of solar energy.

Binary fission Mode of reproduction not involving any sex in which the parent organism, colony, prokaryotic or eukaryotic cell divides into two roughly equal-sized offspring.

Biparentality Condition of having two parents; characteristic of sexually derived offspring.

Blastula Name of animal embryo, always *diploid* (has two sets of chromosomes) because it develops from fusion of egg and sperm as product of fertilization. One fertile egg cell divides by binary fission to form an immature animal that at this blastula stage often looks like a hollow ball; defining characteristic of members of the animal kingdom.

Chemical gradient Environmental or physiological state in which a chemical is distributed at increasing or decreasing concentration. Examples: during oil spills a gradient is established with the highest point (center) at the source (the spill site). Since they move toward places of higher concentration of an attractant chemical, the component cells of a slime aggregate occurs along a chemical gradient.

Chromatid Half chromosome (chromosome after chromatin replication) which is seen in mitosis when one chromatid travels ("segregates") to each end (pole) of the cell. As the one cell becomes two the chromatids become chromosomes (see Plate 23).

Chromatin Easily stained material of which chromosomes are made, i.e., DNA and various histone and nonhistone proteins.

Chromosome Gene (DNA)-bearing structure made of chromatin, usually visible only during mitotic or meiotic nuclear division in stained cells or cells with large chromosomes.

Closed system In the branch of physical science called thermodynamics a bounded study region closed to the flux in and out of matter but not necessarily of energy. The Earth-Sun system is nearly entirely closed to matter (except for meteorites and cosmic particles) but it is entirely open to energy (sunlight and other radiation).

Conjugation Copulation, mating cell contact between complementary genders that results in genetic recombination. In bacteria (prokaryotes) the cells contact at the site through which occurs transmission of genetic material from donor to recipient. (Other forms of prokaryotic sex are not conjugations.) In eukaryotes conjugation refers to isogamous matings: the mating of gametes, gamete nuclei or organisms of equal size and/or form.

Convergence Convergent or parallel evolution. The independent development of similar structures or behaviors in populations that are not directly related, but have lived under the same selective pressures (e.g., the evolution of roughly similar body plans in sharks and dolphins).

Cytocide Unprogrammed cell death due to trauma; membranes rupture and cell contents leak to the environment.

Diploidy Condition of eukaryotic cells in which the nucleus contains two complete sets of chromosomes, that is the nuclei are diploid. Abbreviated 2n to be distinguished from haploidy (one set, 1n). See *haploid*.

DNA Deoxyribonucleic acid, long-chain molecule. The carbon-hydrogen nitrogen-oxygen-phosphorus containing double helix making up the genes of all cells and thus all organism.

Embryonic stem cell Initial cell; cell giving rise by division to identifiable progeny cells. Totipotent cells of animal early embryos can develop into blood or eye cells, for example, depending on their location. The cell at the tip in plants is an embryonic stem cell whose offspring cells can form specific tissue layers such as plant "skin" (epidermis) or pulp (pith).

Endosymbiosis Ecological term referring to a physical association of partners; the condition of one organism living inside another. Describes either within-cell (endocytobiosis such as the bacteria that became the mitochondria or chloroplasts) or inside but between-cell positions (such as the termite protoctists or the root-fungal associations).

Entropy Thermodynamic term which expresses the degree of disorder of a system. The universe as a whole, and most systems within it, tend toward an ever-increasing level of entropy. Although both are made of the same chemical substance (two parts hydrogen to one part oxygen of water, H_2O) more entropy is associated with steam or liquid water than with snow crystals.

Eukaryote Nucleated cell or organism.

Evapotranspiration Evaporation of water from the leaf pores of plants causes a negative pressure which draws water up through the root system released as vapor to the atmosphere..

Fertilization Fusion of the two sex (haploid) cells either gametes (entire cells) or the chromosome-containing parts of gametes (the nuclei) to form the diploid nucleus (fungi, many protoctists), the fertile egg (animals, many plants and protoctists) or the embryo sac (flowering plants). The resultant diploid nucleus in a larger cell, a structure called the zygote, is the most common product of fertilization.

Flagellum Cell whip. Term still often used for undulipodium but should be restricted to the bacterial (prokaryotic) extracellular structure entirely different from the "eukaryotic flagellum." The relatively rigid rod-shaped flagellum, composed of shaft protein called flagellin, moves by rotation at its base. A series of 4 or 5 proteins form rings visible in the cell membrane when seen by electron microscopy.

Gametes Sex cells, e.g., sperm and eggs. In humans and other animals, gametes are the products of *meiosis* (see below); hence they carry half the chromosome complement of other cells. Fusion of sperm with egg (*fertilization*) restores the full chromosome set in the *zygote* (product of fertilization).

Gender Differences between any two complementary organisms that render them capable of mating. Organisms of different gender potentially mate whereas those of the same gender can not mate to form fertile offspring. Many species include healthy organisms of hundreds, even thousands of genders. In some, gender (mating type) differences are determined by tiny changes: specific genes and proteins on the surface of mushroom threads (hyphae) give over 50,000 different genders in the common fungus, *Schizophyllum*. The bewildering series of genders in ciliates that depend on tiny chemical genetic differences in undulipodial surface proteins (ciliary antigens) give rise to genders that may change on a daily cycle. Maleness and femaleness commonly associated with fertilization by anisogamy where male individuals produce many small, swimming gametes (i.e. sperm) while females produce fewer, larger, food-storing gametes (i.e. eggs) is just one of many natural systems.

Genotype Genetic make-up of an organism with respect to specific traits, in contrast to the physical manifestation of those traits (phenotype).

Haploidy Condition of eukaryotes (their cells, tissues or organisms) in which the nucleus contains but a single complete set of chromosomes, that is, the nuclei are haploid. Abbreviated 1n to be distinguished from diploidy (two sets, 2n). See *diploid*. Haploidy is reestablished by meiosis whereas diploidy is by fertilization.

Hayflick number In tissue culture, usually mammalian, the maximum number of cell divisions (generations) that can occur before heritable changes occur (such as mutations including chromosome losses) or death ensues.

Heterotrophy (n. heterotroph; adj. heterotrophic) Mode of nutrition in which organisms obtain carbon from carbon-hydrogen containing substances (such as proteins, sugars and fats themselves derived from chemo- or photosynthesis) but not from carbon dioxide or carbon monoxide. Examples of heterotrophs include osmotrophs (absorptive feeders such as most bacteria and fungi, insectivores, herbivores, carnivores and many others).

Heterozygosity (n. heterozygote; adj. heterozygous) Hybrid genetic condition in diploid organisms that come from genetically different parents. Such heterozygotes have one or more chromosome pairs which bear different alleles (states) of a gene or genes; see *homozygosity*.

Homozygosity (n. homozygote; adj. homozygous) Inbred genetic condition in diploid organisms that come from genetically similar parents. Such homozygotes have one or more chromosome pairs which bear the same allele (state) of a gene or genes; see *heterozygosity*.

Hypersex Permanent symbiogenesis: the origin of new cells, organs or organisms by irreversible physical association between different kinds of live organisms. Hypersex is the process that led to the nucleated, swimming, oxygen-breathing algal cells by associations between different bacteria: some fermenting, some swimming, some oxygen-respiring and others photosynthesizing. Hypersex is a mechanism of evolutionary change in which former symbionts integrate entirely and a formerly loose association becomes permanent. See *symbiogenesis*.

Isogamy Sex cells (gametes) requiring fertilization for continued development that are the same as each other in size or form (as in the two equal gametes of *Chlamydomonas*). The mating (pairing) of gametes that are alike in morphology and size to complementary mating types that they resemble (isogametes) occurs commonly in protoctists and fungi. See *anisogamy*.

Isolated system In the branch of physical science called thermodynamics, an isolated, bounded study region that is closed to the flux in and out of both matter and energy. A sealed vacuum flask of hot coffee is a commonplace, though imperfect, attempt to produce an isolated system.

Kinetochore Centromere. The motor on the chromosomes that moves them along the microtubules to the poles in cell division (mitosis and meiosis). DNA/protein structure usu. located at a constricted region of a chromosome, that holds chromatids together; also the site of attachment of spindle fibers during *mitosis* and *meiosis*.

Macrophagy Mode of heterotrophic nutrition in which organisms (*macrophages*) feed on food particles large with respect to their own size. Macrophage also refers to amoeboid white blood cells, that engulf pathogens as part of the mammalian immune response.

Magnetotaxis (adj. magnetotactic) Directed swimming of a live organism in a magnetic field toward a magnetic pole (e.g. south- or north-seeking; as in magnetite-containing bacteria or *Chlamydomonas*).

Meiosis Cell division with halving of chromosome number, e.g. one or two divisions of a diploid parent cell which give rise to haploid offspring cells. In some organisms (e.g., animals), meiosis in diploid body cells precedes the formation of haploid *gametes* (see above). In other organisms (e.g., fungi), the transient diploid *zygote* (see below) undergoes meiosis immediately after it forms, giving rise to haploid body nuclei. Chromosome pairing may be seen during most meioses. Meiosis as an essential component of fusion sex cycles is found in all plants and animals, and all sexual fungi. Meiotic sex is present in many, but by no means all, protoctists (see Plate 16).

Metabolism Chemical reactions of energy and heat-generating living matter including DNA synthesis (gene replication) and protein production (RNA and protein synthesis) that underlie autopoiesis. Waste products accrue as energy is generated.

Microtubule (adj. microtubular) Slender, hollow, proteinaceous structure essential to cell motility of eukaryotes. Microtubules are of varying lengths but usu. invariant in diameter at 24-25 nm; substructure of mitotic spindles, undulipodia and many other intracellular structures.

Mitosis Cell division with maintenance of chromosome number, e.g. division of a haploid or diploid parent cell gives rise to genetically similar offspring cells. Four stages are usually recognizable: prophase, in which centrioles divide and chromosomes condense; metaphase, in which chromosomes move and align at the equatorial plane of the nucleus; anaphase, in which the chromosomes separate at their *kinetochores* and move to opposite poles; and telophase, in which the chromosomes return to their extended state and the entire cell divides.

Mitotic spindle Transient microtubular structure that forms in dividing eukaryotic cells and is responsible for chromosome movement to opposite poles during *anaphase* (see *mitosis*).

Morphospecies Species that can be distinguished on the basis of form (morphology) and are therefore named differently.

Mutation Heritable DNA change; ultimate source of genetic variation in evolving populations.

Necrosis Death of cells, a piece of tissue or an organ in an otherwise living organism.

Neoteny Retention of juvenile features in sexually mature adult animals. *Examples:* Adult human skulls are more similar to those of our infants than adult other primate species skulls are to theirs. Gorilla skulls change radically during maturation. The testes and ovaries of the adult (terrestrial) axolotl salamander develop in the juvenile (swimming) stage. Neoteny often seems to correlate with relatively recent evolution of a species.

Nucleoid Organelle of bacteria, the DNA-containing structure of prokaryotes, not bounded by a membrane. Synonymous with genophore.

Nucleus The organelle of eukaryotes that is universally present. This membrane-bounded, usually spherical, DNA-containing body is the site of both DNA and RHA synthesis. Chromatin is organized into chromosomes within the nucleus.

Open system In the branch of physical science called thermodynamics, a bounded study region open to the flux in and out of both matter and energy. Living systems, e.g., the semipermeable cell, are open.

Organelle Visible structure inside a cell like an undulipodium, mitochondrion, nucleus or chloroplast.

Parthenogenesis Virgin Birth. Developmental state in which unfertilized eggs develop into offspring at a given life history stage; characteristic of many animal and protoctist symbionts that need not mate in order to produce offspring from eggs. Animals formed by a single parent, the mother, with no intervention by sperm.

Phagocytosis (adj. phagocytotic, phagocytic) Mode of heterotrophic nutrition and immune defense involving ingestion, by a cell, of solid particles. Characteristic of amebas and macrophages (white blood cells) whose pseudopods (cytoplasmic "blebs") flow over and engulf invading bacteria or other blood-borne particles.

Phenotype Physical make-up of an organism with respect to specific traits and/or set of environmental conditions in contrast to the underlying genetic basis of those traits (*genotype*).

Pheromone A kind of allelochemical, a chemical substance that when released into an organism's environment, influences the behavior or development of others of the same species. If produced by one gender and responded to by another, the substance is called a sex pheromone.

Photon A minimal unit of light, a particle (quantum) of electromagnetic radiation

Photosynthesis Photoautotrophy. Mode of nutrition in which light provides the source of energy. A photoautotrophic organism uses light energy to make cell materials from inorganic compounds (carbon dioxide, nitrogen salts, phosphates).

Pineal gland The pineal organ, so named because in mammals it is shaped like a pine cone; a protrusion from the base of the brain that produces the hormone melatonin. Melatonin production is associated with the diurnal cycle; in some mammals, melatonin levels have been shown to affect sperm and egg production, so that reproduction goes "on hold" in response to long winter nights.

Pole (see *mitotic spindle*) Ends where mitotic spindle fibers during mitosis or meiosis come to a point in the dividing cell; in many organisms, location of centrioles.

Prokaryote Bacterial cell or organism.

Protein Long-chain molecules made of carbon, hydrogen, nitrogen, oxygen and sulfur that comprise most of the dry weight of all organisms. Movement, acceleration of chemical reactions, maintenance of salt balance and myriad of other functions are performed by proteins. Most cells incessantly produce over a thousand different kinds of protein essential to the livingness of matter.

Replication Process that augments by copying the number of DNA or RNA molecules. Molecular duplication process.

Reproduction Any process that leads to the increase in the number of living individuals. A single parent is capable of reproduction (by binary fission, budding and other processes that do not involve sex) whereas two parents are necessary for sexual reproduction.

RNA Ribonucleic acid, long chain molecule closely related to dna, the nitrogen-oxygen-phosphorus containing single chain molecule that acts as messenger between dna and protein and plays other roles essential to all cells and thus all organisms.

Runaway selection Mating success where genes that produce exaggerated phenotypes and exaggerated mating preferences are selected and retained in a population. The male Irish Elk (*Megaloceros*) had an 11-foot, 100-lb rack of antlers due apparently to its attractiveness to elk females.

Seme Complex trait of identifiable selective advantage, and therefore of evolutionary importance, resulting from evolution of an interacting set of genes. Unit of study by evolutionary biologists (e.g. nitrogen fixation, cell motility, eyes).

Sex Formation of new organism containing genetic material from more than a single parent. Minimally involves uptake of genetic material from solution and DNA recombination by at least one live (autopoietic) entity. Sex also refers to a mode of reproduction that involves the formation of haploid nuclei in eukaryotes (meiosis) and fertilization to form *zygotes*.

Sexual dichromatism A kind of sexual dimorphism where the differences in males and females belonging to the same species are differences in coloration. Common in birds e.g., cardinal, bluejay.

Sexual dimorphism Body size and form, behavior and/or metabolic differences in males and females belonging to the same species. Found in algae, some flowering plants and all mammals.

SOS response Initiated by DNA damage (chemical, radiation etc.), bacterial proteins of DNA repair are activated. The repair process itself is error-prone and can lead to new mutations.

Species Set of named, identified and classified distinguishable organisms that are usually detectable by their characteristic bodies, metabolism and behavior. Members of the same species have most of their genes in common; in animals and plants any member of a given species either does not mate or it remains infertile when it attempts to mate with an organism of other species.

Superordination Acquisition of higher rank, status or value; in biological terms, the tendency of a population to assume the characteristics of a single large organism, particularly through associative behaviors of slime molds, naked mole rats, ants, bees, wasps.

Symbiogenesis The origin of new cells, organs or organisms by physical associations between different kinds of live beings; may be cyclical or permanent. See *hypersex*.

Symbiosis Prolonged physical association between two or more different organisms belonging to different species. Levels of partner integration in symbioses may be behavioral, metabolic, gene product, or genic.

Thermodynamics A quantitative science, the branch of physics that deals with the relationships between heat and other forms of energy.

Totipotency Developmental term referring to propagule or growing cell that is capable of repeating all steps of development and giving rise to new complete organisms or a wide array of different cell types. See *embryonic stem cells*.

Transduction The transfer of small replicons (e.g. viral or plasmid DNA) from an organelle or bacterium to another organelle or bacterium, usually mediated by a virus. Also: Change of energy from one form to another (e.g., light to chemical or mechanical energy to heat).

Transfection Natural genetic change in bacteria and eukaryotic cells in culture induced by uptake of DNA from aqueous medium.

Transgenic organism Organism incorporating one or more genes of foreign origin (transgenes), often expressing phenotype(s) associated with the transgene. Transgenic *Escherichia coli*, genetically engineered bacteria incorporating a human gene, are used for large-scale commercial production of human insulin.

Undulipodium Cilium, sperm tails and all cell protrusions traditionally called "eukaryotic flagella." These moving organelles that sometimes show feeding or sensory functions are composed of at least 200 proteins. In cross section the undulipodium which is covered by the plasma membrane displays a distinctive structure: the $[9(2)+2]$ microtubular array. Since undulipodia contrasts in every way with the much smaller flagellum of bacteria (the prokaryotic motility organelle with which undulipodia should not be confused) the unambiguous term undulipodium is prefered.

Zygote Fertile egg or other diploid (2n) nucleus or cell produced by mating. Although in all cases two haploid nuclei or cells fuse (fertilize) to form zygotes, in animals, plants and some protoctists (those undergoing gametic meiosis) the zygote is destined to develop into a new organism. In fungi and protoctists undergoing zygotic meiosis, the short-lived zygote stage is unstable; meiosis produces haploid nuclei again soon after the zygote is formed .

ACKNOWLEDGMENTS | LIKE THE SEX ACT ITSELF which always requires more than a single live being, the making of a book is never an act of the authors alone. Inspired and aided by the tireless efforts of Peter N. Nevraumont, only his generosity and that of other friends made **What is Sex?** possible. We especially thank Lois Brynes, Michael Chapman and Michael Dolan for their role in manuscript preparation and fact-finding. Robin Kolnicki gave us many excellent research leads and technical assistance. Tonio Sagan provided encouragement from the wings. We thank Howard Bloom for sharing his work on group selection and permission to use his words on cybersex. Steven Shaviro magnanimously allowed us to quote from his work on post modernism. Thanks to Terry Bristol for putting us up in Oregon and to Matt Rotchford for driving to the "Castle of Chaos," the Albany, Oregon home of Jerry Andrus, whom we thank for permission to publish his wonderful parabox illusion, which appears on page 174. Help also came from many students, colleagues and friends including Ricardo Guerrero, Aaron Haselton, Jeremy Jorgensen, Kelly McKinney, Seth! Leary, Lorraine Olendzenski, Donna Reppard and Eric Schneider.

We are grateful to Charlene Forest of Brooklyn College who mated and photographed her *Chlamydomonas* for us at the last minute of book production.

Special thanks are due to artist Kathryn Delisle who created and corrected her clear illustrations from our rough notes. And to José Conde for his wonderful design.

At Nevraumont Publishing, we thank Ann J. Perrini, President, and Simone Nevraumont, Assistant Editor, for their many contributions to making this book a success. And Katherine Hasal for her hard work in copy editing the manuscript.

Financial support for our contribution to the underlying scientific research at the Margulis laboratory is primarily from NASA. The Richard Lounsbery Foundation and the College of Natural Science and Mathematics at the University of Massachusetts Amherst have also helped sponsor our work.

index

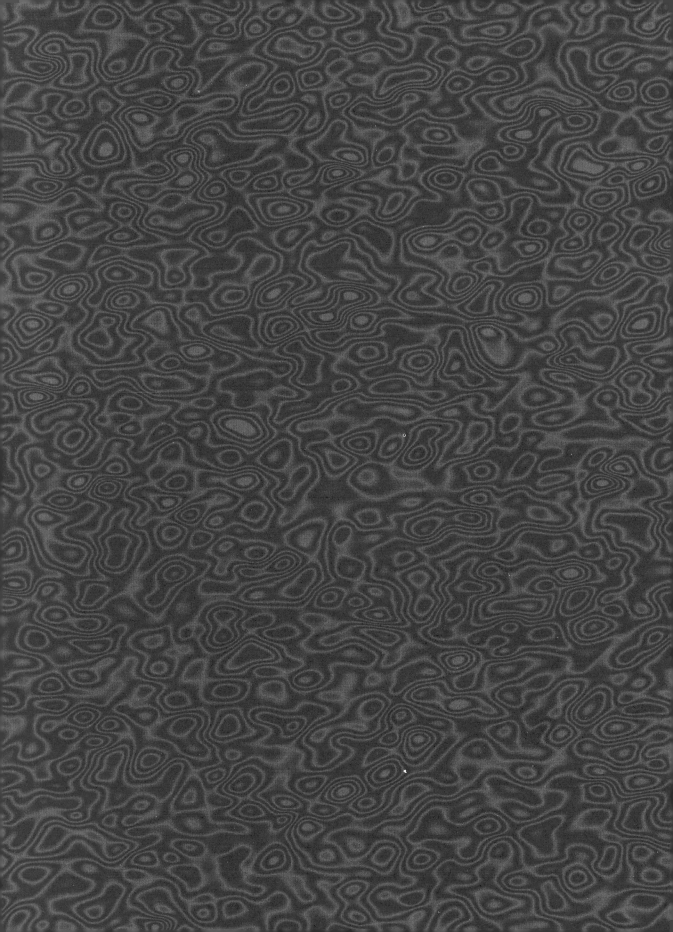